Synthesis Lectures on Engineering, Science, and Technology

The focus of this series is general topics, and applications about, and for, engineers and scientists on a wide array of applications, methods and advances. Most titles cover subjects such as professional development, education, and study skills, as well as basic introductory undergraduate material and other topics appropriate for a broader and less technical audience.

Gürol Yıldırım

Hydraulic Principles and Design Concepts for Submain Units with Multiple Outlet Pipelines

New Analytical Techniques with Engineering Applications

Springer

Gürol Yıldırım
Professor of Hydraulic and Civil Engineering
Giresun University Rectorate, Vice Rector
Giresun, Türkiye

ISSN 2690-0300 ISSN 2690-0327 (electronic)
Synthesis Lectures on Engineering, Science, and Technology
ISBN 978-3-031-32497-0 ISBN 978-3-031-32495-6 (eBook)
https://doi.org/10.1007/978-3-031-32495-6

This Springer imprint is published by the registered company Springer Nature Switzerland AG
The registered company address is: Gewerbestrasse 11, 6330 Cham, Switzerland

Preface

The book is suitable as a useful reference for senior-level undergraduate and graduate students, in irrigation or civil and environmental engineering, practicing engineers and researchers. The students are expected to have had a fluid mechanics course before studying this book. In line with this concept "hydraulic design of multiple outlet pipes", although a large number of technical papers are readily available, the user-friendly nomographs or the books which could better clarify the background derivative procedures of the used techniques, and demonstrate their practical applications to cover various types of design problems, is very limited.

The main objective of this book is to provide a state-of-the-art review of recent analytical developments in multi-outlet pipe flow hydraulics and alternative hydraulic design concepts. For this purpose, the book presents simple, but sufficiently accurate, analytical equations, which can be applied directly without needing any numerical technique to achieve whole hydraulic computations. The analytical procedures provided in the book give an opportunity for the designers to better understand the basic hydraulic principles of multi-outlet pipe flow, and demonstrate their applicability and efficiency on design problems of multi-outlet submain lines, covering various design configurations.

The field of application of the book is efficient for manifold flow applications. First, it mainly deals with the hydraulics of multi-outlet submain lines laid on low-pressurized micro/sprinkler irrigation systems, and also covers areas such as assessment for leaky pipes in branched networks of water distribution systems, multi-port diffuser manifolds for sea outfalls, oil pipelines or industrial multiple outlet piping systems.

Brief Summary of Book Chapters' Organization

The book mainly covers five chapters. Each chapter of the book is developed in a systematic manner from the first basic hydraulic principles presented in Chap. 1. A very strong emphasis is given to practical design applications, and unnecessary theoretical details have been avoided as much as possible. However, the various chapters have been written so that they can be read individually. Sometimes, however, other parts of the book had to be referred to in order to avoid duplication. This has been done in such a manner that only the section referred to may be read and not the whole chapter. This mode of presentation will allow practicing engineers to read only those parts of the book that are of immediate interest, and will allow teachers to select the chapters most relevant to their courses. For convenience, design curves are also appended to the text in some chapters, for quick computations during the preliminary design stages.

The Sequence of Presentation is as Follows

Chapter 1 presents an overview of the classification (Class A, B, C) covering various types of multi-outlet pipe design problems with respect to the main required parameters. Fundamental concepts of fluid mechanics as applied to multi-outlet pipe flow hydraulics are introduced. As governing equations, the conservation laws are revisited, and the equations of continuity, momentum and energy are derived. First, five types of pressure head (energy) profiles (Type-I, Type-IIa, Type-IIb, Type-IIc and Type-III), which covers various uniform line slope situations, are clearly identified. This chapter presents a new analytical "Energy-Gradient Ratio (EGR)" approach to identify first, which type of pressure profile occurs for a given uniform design slope when other hydraulic variables are initially known, then comprehensively evaluate its hydraulic characteristics along the line. It is revealed in this chapter, among all types of pressure profiles, the Type-IIb profile is considered as the optimal (or ideal) pressure profile which can produce the minimum pressure head difference which is regarded as "*the most efficient pressure profile*" or "*ideal hydraulic design*" in the literature. At the end of this chapter, a comprehensive examination of the proposed mathematical formulations is presented for all performed simulations.

In Chap. 2 to analyze and better understand the main concern "multi-outlet pipe flow hydraulics", a new analytical procedure "Revised Energy-Grade Line (REGL) Method", which is based on the "non-uniform outflow" concept, is developed using the governing equations presented in Chap. 2, instead of the Energy-Grade-Line (EGL) method based on the traditional "uniform outflow" approach.

In Chap. 3, the REGL method is essentially improved to evaluate the relative effects or contributions of minor (local) energy losses to total energy losses, and the required hydraulic design parameters, along the submain line. The design curves are also appended to the text, for rapid evaluation alternative design cases (with or without minor losses), without using the complete analytical derivations.

Chapter 4 presents analytical design concepts which improve the REGL method previously defined in Chap. 3, to directly determine multi-outlet pipeline geometric characteristics. For any desired uniformity level, the procedure provides an opportunity to evaluate the influence of local energy loss, as well as the influence of different uniform line slopes on the pipe geometric characteristics (pipe size and length), and on the corresponding hydraulic variables (operating inlet pressure head, downstream end pressure head and total energy losses).

Chapter 5 presents the computer-aided Backward Stepwise Procedure (BSP), to evaluate the relative effects of all energy loss components of total energy losses along the lateral lines. In that procedure, the friction losses due to pipe and emitters, and the local losses due to insertions of different types of on-line and in-line emitters, are clearly analyzed, and the results are compared/verified with those of the recent experimental procedures.

Giresun, Türkiye

Gürol Yıldırım
PhD. Professor of Hydraulic and Civil Engineering

Acknowledgements

As book's authors, we gratefully appreciate our colleagues for their skill in reviewing, editing and possible contributions to some book sections, in advance. Their careful attention to detail has resulted in a better presentation of this authored book.

The book author wants to express his acknowledges to Prof. Vijay P. Singh at Texas A&M University, Emeritus Prof. Necati Ağıralioğlu from International Antalya Science University, and Distinguished Emeritus Prof. Zekai Şen from Istanbul Medipol University, for their high encouragement on preparing this book, many scientific activities, projects and so on. The Scientific and Technological Research Council of Turkey (TUBITAK) is acknowledged for supporting through the fellowships and grant program during the studies at the Texas A&M University/BAEN. We also thank the publishers, potential readers and reviewers for providing constructive reviews and criticism on the content of the book, in advance.

I'm also sincerely and kindly grateful to my wife Ms. Semra Yıldırım, my daughter Mihrimah Ruveyda Yıldırım and my son Dr. Mücahit Yıldırım, for their patience, support and continuous interest in preparing this book. My dearest family given me as I took time from our activities to find time to pursue this long-period project.

Thanks to all of you!

2023

Gürol Yıldırım, Ph.D.
Head of Hydraulics Division
Full Professor of Civil Engineering
Giresun University
Giresun, Türkiye

Contents

About the Author

Gürol Yıldırım is full Professor of Civil Engineering, Head of Hydraulics Division, Civil Engineering Department and Vice Rector of Giresun University, Turkey.

Born in Konya, Turkey, he received his Ph.D. degree from Hydraulic Engineering Division at Istanbul Technical University, Civil Engineering Faculty, in 2007. During his doctorate at Istanbul Technical University, he taught courses in fluid mechanics, hydraulics, water resources, hydrology, water supply and environmental sanitation; river hydraulics and restoration, flood control and its hydrology, manifold and diffuser hydraulics in water distribution networks. He has been at Texas A&M University (TAMU) Biological and Agricultural Engineering Department (BAEN) for post-doctorate research and joined in collaborative projects with Dr. Vijay P. Singh. Dr. Yıldırım received his associate professorship degree in 2011, and professorship degree in 2016, in Civil and Hydraulic Engineering Department.

His research is mainly conducted on the hydraulic and economic analysis and design concepts on water distribution networks, pipe flow hydraulics analysis, i.e. low-volume and low-pressurized irrigation systems, turbulent and resistance law phenomena pipe turbulent flows, friction factor computation methods using truly analytical, numerical and semi-theoretical approaches and new soft-computing techniques, and commercial pipeline design. He has published full-paper technical articles, technical notes, discussions and closures, more than 100, in respective publishers; reviewed numerous technical papers; published book chapters, technical reports in several major projects.

Dr. Yıldırım received "the Best Discussion Award", from American Society of Civil Engineers (ASCE) at the Environmental Water Resources Institute (EWRI) Conference, in Honolulu/Hawaii, 2008. He is a member of the American Society of Civil Engineers (ASCE) and also Member of Science and Advisory Board of the Journal of Environment, Urban and Climate, published under the Republic of Turkey, Ministry of Environment, Urbanization and Climate Change. He lives with his wife Semra, his daughter Mihrimah Ruveyda and his son Mücahit in Giresun/Turkey.

Symbols

A	Cross-sectional area of the multiple outlets pipe [L^2]
A, B	Empirical parameters for the Churchill friction coefficient formula [Eq. (9b)]
C	Combined friction coefficient for the Darcy-Weisbach formula, $C = 8c_1(\upsilon/4)^{2-m}/(g\pi^m)$
CV_H	Coefficient of variation of pressure heads, relationship between standard deviation and mean (%)
CV_q	Coefficient of variation of discharges, relationship between standard deviation and mean (%)
c_1	Constant for the particular friction coefficient formula used
D	Internal diameter of the pipe [L]
DEV_l	Difference between pressure head and average pressure head at any distance l from the pipeline inlet [L]
DU_{LQ}	Lower-quarter distribution uniformity coefficient (%)
dl	Strip of length considered at any length from the pipeline inlet [L]
dS	Area of the strip of length dl, $dS = H_f(l)dl$ [L^2]
F	Traditional friction correction factor for infinite number of outlets
F_C	Christiansen's friction correction factor for finite number of outlets depends upon the inlet spacing ratio
F_{SW}	Stepwise friction correction factor taking into account effect of the number of outlets
f	Friction coefficient for the Darcy-Weisbach formula
g	Acceleration due to gravity [LT^{-2}]
H	Operating pressure head at the head of the outlet [L]
$\overline{H}$	Required average outlet pressure head [L]
$\overline{H}_{0.5}$	Average pressure head along the upper half segment of the pipeline [L]
H(l)	Pressure head at a given length l from the pipeline inlet [L]
$H_V(l)$	Velocity head at a given length l measured from the pipeline inlet [L]
H_{VI}	Velocity head at the head end of the pipeline, $H_{VI} = 8(N\overline{q})^2/\pi^2 g D^4$ [L]
H_I	Pressure head at the pipeline inlet [L]

H_1	Upstream pressure head at the first outlet near the pipeline inlet [L]
H_N	Downstream pressure head at the first outlet from the closed end of the pipeline [L]
H_{max}	Maximum operating outlet pressure head [L]
H_{min}	Minimum operating outlet pressure head [L]
H_{var}	Variance of operating outlet pressure head [L]
$H_f(l)$	Friction head loss per unit weight of fluid at any distance l from the pipeline inlet [L]
$H_{f(L)}$	Total friction drop at the downstream closed end for the entire length of the pipeline, $H_{f(L)} = FH_{f0}$ [L]
$H_{f(Lo)}$	Friction drop for length of the pipe section L_o between the first and last outlets [L]
H_{f0}	Total friction drop in a similar pipeline without outlets, $H_{f0} = CQ_I^m/D^n$; $H_{f0} = f(L/D)H_{VI}$ [L]
$\overline{H}_f$	Average head loss due to friction along the pipeline [L]
$\overline{H}_{f0.5}$	Average head loss due to friction along the upper half segment of the pipeline [L]
i	Dimensionless variable of the original distance l which is measured from the pipeline inlet, $i = l/L$
i_{max}	Dimensionless distance where the maximum pressure head (or outflow) occurs
i_{min}	Dimensionless distance where the minimum pressure head (or outflow) occurs
K	Integer represents the consecutive order of outlets
k	Coefficient in outlet discharge formula $[L^{3-x}T^{-1}]$
L	Full length of the pipeline including inlet spacing from the inlet to the first outlet [L]
L_0	Length of the distance between the first and last outlets except for the inlet spacing [L]
l	Space coordinate; original distance which is measured from the multiple outlet pipeline inlet [L]
l_{max}	Location where the maximum pressure head (or outflow) occurs [L]
l_{min}	Location where the minimum pressure head (or outflow) occurs [L]
$l_{fully\text{-}turbulent}$	Length of the upstream segment in which the fully-turbulent flow prevails along its length [L]
$l_{turbulent}$	Length of the upstream segment in which the turbulent flow prevails along its length [L]
$l_{transition}$	Length of the downstream transition zone in which the indeterminate flow prevails along its length [L]
$l_{laminar}$	Length of the downstream segment in which the laminar flow prevails along its length [L]

m	Velocity (or flow rate) exponent for the particular friction head loss formula used
m_β	Improved value of the velocity exponent for the nonuniform outflow, $m_\beta = \beta$ m
N	Total number of equally spaced operating outlets for the entire pipeline
n	Diameter exponent for the particular friction head loss formula used
p	Slope indicator takes value 1 for uphill slope, −1 for downhill slope and 0 for zero slope
Q	Flow rate that passes through the pipeline [L^3T^{-1}]
Q_I	Total inflow rate for the entire length of the pipeline, $Q_I = N\overline{q}$ [L^3T^{-1}]
Q(l)	Total flow rate that passes through the pipe section located at any distance l from the pipeline inlet [L^3T^{-1}]
q	Discharge through the individual outlet [L^3T^{-1}]
q_i	Outflow per unit length of the pipeline [L^3T^{-1}]
q(l)	Continuous outflow function at any distance l from the pipeline inlet [L^3T^{-1}]
$\overline{q}$	Required average outlet discharge [L^3T^{-1}]
q_{max}	Maximum outflow from an individual outlet [L^3T^{-1}]
q_{min}	Minimum outflow from an individual outlet [L^3T^{-1}]
$q_{max}-q_{min}$	Maximum difference in the extreme outflows along the pipeline [L^3T^{-1}]
R	Reynolds number for pipe flow
R_I	Reynolds number at the inlet of the pipeline, $R_I = (4N\overline{q}/\pi D\nu)$
R_i	Friction drop ratio, $R_i = H_f(l)/H_{f(L)} = \left[1-(1-i)^{m+1}\right]$ (%)
R(l)	Reynolds number for any distance l from the pipeline inlet
s	Common spacing between consecutive operating outlets [L]
s_I	Inlet spacing from the inlet to the first outlet, $s_I = \mu$ s [L]
$S_V(l)$	Velocity head gradient at a given distance l
$S_f(l)$	Friction slope or the slope of the energy grade line at any distance l from the pipeline inlet
S_0	Uniform slope along the pipeline
t	Time coordinate [T]
U_C	Christiansen's uniformity coefficient (%)
$\overline{V}$	Average velocity of flow through the pipeline [LT^{-1}]
$\overline{V}(l)$	Average velocity of flow at a given section of length l measured from the pipeline inlet [LT^{-1}]
$\overline{V}_I$	Average velocity of flow at the pipeline inlet [LT^{-1}]
x	Outlet discharge exponent
ΔH	Maximum allowable difference in the outlet operating pressure heads along the pipeline [L]
Δq	Maximum allowable difference in the outflows along the pipeline [L^3T^{-1}]

ΔS	Total head loss or gain due to elevation at the closed end of the pipeline, $\Delta S = S_0 L$ [L]
$\overline{\Delta S}$	Average head loss or gain due to elevation for the entire length of the pipeline, $\overline{\Delta S} = \Delta S/2$ [L]
$\overline{\Delta S}_{0.5}$	Average head loss or gain due to elevation for the upper half segment of the pipeline, $\overline{\Delta S}_{0.5} = 0.75\Delta S$ [L]
δ_H	Percentage of the maximum allowable difference in outlet pressure heads to the average value (%)
δ_q	Uniformity criterion for the emitter outflow variation (%)
ε	Equivalent roughness height of the internal pipe surface [L]
ν	Kinematic viscosity of water [L^2T^{-1}]
μ	Inlet spacing ratio, $\mu = s_I/s$ $(0 \leq \mu \leq 1)$
β	Nonuniform outflow exponent, $\beta = 1 - 3.322x \log(\overline{H}_{0.5}/\overline{H})$
λ	Formulation constant, $\lambda = (1 - 0.5^{m+1})/(m+2)$
θ	Length ratio, $\theta = L_0/L$

List of Figures

List of Tables

1 Review of Fundementals in Hydraulic Engineering Perspective: Hydraulics Principles, Governing Equations, Design Criteria, Types of Design Problems, Hydraulics and Energy-Gradient Line Profiles for Uniformly Sloping Multiple Outlets Pipelines

Introduction

Numerous researchers have worked on the hydraulic analysis of multiple outlets pipeline networks [1–6]. In this classification of the design problems, determination of the actual pressure head (or outflow) profile along the line is greatest important concern for hydraulic design [7–11]. The energy-gradient line (EGL) method is traditionally used for the determination of lateral pressure head and outflow profiles [12–14]. Application of the analytical EGL or Revised Energy-Grade Line (R-EGL) [15–17] method is very easy; however, it has a limitation of the application region due to the constraints in the outlet flow regime and outflow variation along the line.

Recently, an improved EGL method has been presented [18] to solve hydraulic design problems of various types of multi-outlet submain lines in different flow regimes and uniform line slope cases. In the analytical development, the improved energy-grade line is determined based on the average friction drop ratio approach with a simple exponential function, to express the nonuniform outflow concept [8, 19]. The solution technique of this analytical procedure is simple and direct, and sufficiently accurate in comparison with the numerical stepwise procedures [20–22].

The operating inlet pressure head, H_{0I}, is a main hydraulic component for the proper design and evaluation of pressure head distribution along the line [22]. This head can be achieved in a stepwise manner, either starting from the required pressure head at the downstream closed end, H_d, working back to the inlet upstream direction, by computing the friction head loss in each pipe segment between successive outlets (BSP: backward-step procedure) [23, 24]; or starting from the inlet pressure head, H_{0I}, computing forward in the downstream direction (FSP: forward-step procedure) [20, 21]. In this procedure, H_{0I}

G. Yıldırım, *Hydraulic Principles and Design Concepts for Submain Units with Multiple Outlet Pipelines*, Synthesis Lectures on Engineering, Science, and Technology,
https://doi.org/10.1007/978-3-031-32495-6_1

is determined by adding a reasonable head increment to the required average outlet pressure head, ensuring initial and boundary conditions to be based on the design algorithm discussed earlier [8, 9].

In order to avoid the stepwise computer-aided design procedures [8, 22–24], the authors [6, 25] presented a graphical relationship among the required average outlet pressure head, pipe friction loss, and operating inlet pressure head for a sample design example, This procedure is efficient for the design of a multioutlet submain line with single-diameter and two-diameter, as well as multidiameter pipelines.

In conventional design procedures [6, 25], H_{0I} is essentially related to the main hydraulic variables of the energy profile, such as uniform pipe slope; S_0, downstream closed end pressure head, H_d; and/or total friction drop, $h_{f(L)}$, and required average outlet pressure head, $\overline{H}$. However, it may be hard to compute H_{0I}, directly, as $\overline{H}$ and S_0 are known a priori, whereas H_d and/or $h_{f(L)}$ cannot be determined in advance, since the pipe diameter, D, is an additional unknown parameter for the required operating inlet pressure head, H_{0I}, in most usual design cases [26, 27].

To overcome the above problem, the author [7] proposed an alternative computing technique to directly determine H_{0I} which incorporates different uniformity patterns for water application [10, 11]. In this analysis, first, simple mathematical expressions are deduced to relate uniformity parameters; then H_{0I} is simply reformulated by setting a multiplification or enlargement factor α for the required average outlet pressure head, and β^* for the potential energy head change due to uniform line slope. This procedure is simple, direct, and sufficiently accurate to determine the required hydraulic variables along the pressure head profile. However, the applicability of this procedure is only limited for the design case of a Type-I pressure profile for horizontal or uphill slope situation.

The authors [10, 11] have presented a procedure to solve the first and second types of design problems, based on improved Energy Gradient Ratio (EGR) approach which is an improvement on the previous mathematical development, utilizing analytical approaches [10, 11] and which covers different types of pressure head profiles (Type-I, Type-II.a, Type-II.b, Type- II.c and Type-III) under different uniform line slope situations, while determining values of pressure parameters (H_{0I}, H_d, H_{max}, H_{min}, $h_{f(L)}$, ΔH_{max}). The proposed procedure incorporates simple mathematical formulations encompassing different uniformity patterns for water application, such as Christiansen's uniformity (U_C), lower-quarter distribution uniformity (D_U), coefficient of variation of outflows (CV_q), emission uniformity (E_U), and fraction of the required average pressure head (δ_H).

The procedure is simplified by regarding the localized head loss along the pipe line but neglecting the change in kinetic head. The proposed methodology is found to be simple, sufficiently accurate, and can be used efficiently for different types of pressure profiles in different line slope situations. It offers considerable flexibility by directly computing values of the required hydraulic variables along the energy profile, incorporating different uniformity patterns and without requiring any additional hydraulic variables, such as pipe diameter, total inlet discharge, required average outflow, downstream end pressure head, total friction drop or outlet hydraulic characteristics.

Classification on Design Problems of Multiple Outlets Pipelines

Analysis of the hydraulic design of multi-outlet submain lines is important for the proper performance of water supply systems. The design procedure of a multi-outlet plastic pipe often requires determination of the nominal pipe diameter, D_n or the pipe length, L, as well as the required operating inlet pressure head, H_I, total and/or pipe friction losses along the pipe, $h_{T(L)}$ and/or $h_{f(L)}$, ensuring the desired level of hydraulic uniformity criteria (allowable pressure head variation, f_H, Christiansen's uniformity coefficient, U_C or lower-quarter distribution uniformity, D_U and coefficient of variation of outlet discharge, CV_q), for a given pipe slope, S_0, with the remaining initially known design variables; such as total inflow rate Q, average outlet discharge and pressure head ($\overline{q}$ and $\overline{H}$), outlet discharge exponent, x and distance between successive outlets, s. Design engineers are often faced with the following three types of problems in most usual design cases.

Class A Problems: Determination of Energy-Gradient Line (EGL)

In the first type of problem (energy profile determination), for the given design values of S_0 and $\overline{H}$, the required hydraulic variables concerning the pressure head profile [the operating inlet pressure head (H_{0I}), downstream closed end pressure head (H_d), extreme pressure heads (H_{max} and H_{min}) with their locations along the line (i_{max}, i_{min}; i: percentage of length, l is a length from the inlet in m, and L is the total length of the submain line in m), and the total energy drop due to friction $h_{f(L)}$], can be determined ensuring the desired level of water application uniformity as well as the allowable pressure head variation along the line [28–30].

Class B Problems: Determination of Type of Pressure Profile, Operating Inlet Pressure Head and Related Pressure Parameters

In the second type of problem (type of pressure profile determination), when the proper values of water application uniformity parameters (U_C, D_U, CV_q, E_U) are previously known, the operating inlet pressure head (H_{0I}) and the related pressure parameters along the energy-gradient line (EGL) can be directly determined depending upon these uniformity parameters together with the average outlet pressure head ($\overline{H}$) and given design slope (S_0) [10, 11, 28–31].

Class C Problems: Determination of Pipeline Geometric Characteristics (Design of Pipe Diameter and Length)

In the third type of problem (direct sizing procedure), the required parameters are the pipe diameter (D) or length (L) for a given design slope (S_0), the desired level of water application uniformity, and the required average outlet pressure head ($\overline{H}$), with the remaining variables as design variables [32–44].

Hydraulic Principles on Multi-outlets Pipe Flow

Multiple outlets pipelines (manifolds and laterals) in a branched submain unit that are the main part of any water distribution system (sprinkle or trickle) [7, 8] can be regarded as a hydraulic structure whose design is limited by the operating inlet pressure head and by the desired level of water application uniformity that is affected by the allowable pressure head variation and the total energy loss along the pipeline, the field topography, as well as the outlet hydraulic characteristics. These criteria are essentially based on the hydraulic design of water distribution system either to be accepted or rejected. Analysis of the hydraulic design of multi-outlet submain lines is important for the proper performance of water distribution systems. Design of this system depends upon a good understanding of multi-outlet pipe flow hydraulics and outlet (sprinklers or emitters) hydraulic characteristics.

The Basic (Governing) Equations

Generally, outlets (sprinklers or emitters) are usually identical and are installed at an equal spacing on the lateral line. The flow characteristics of outlets are typically described by a function of the form:

$$q_n = cH_n^x \tag{1.1}$$

in which q_n = outflow from an individual outlet; H_n = pressure head in the lateral pipe at the outlet under consideration; and c = outlet coefficient that accounts for areal and discharge effects and makes the units correct. For a lateral, the spacing between successive individual outlets (s), the outlet coefficient (c), and the cross-sectional area of the lateral pipe (A) are taken as constants.

The outlet discharge exponent x characterizes the flow regime and outlet type. Values of x should range from 0 for a pressure-compensating emitter to 1.0 for an emitter in a laminar flow regime. The value of the discharge exponent should be close to 0.5 for emitters operating in a turbulent flow regime. The higher the value of the emitter discharge exponent, the greater the degree of care required to maintain the proper pressure distribution along the lateral for the same uniformity of application. Further knowledge can be obtained [5].

Assuming that the outflow varies continuously in space along the lateral (the number of emitters is sufficiently large), the outflow per unit length (q) can then be described by:

$$q = \left(\frac{c}{s}\right) H_n^x \tag{1.2}$$

Conservation of mass is the most basic principle. If continuity is preserved along lateral, the conservation of mass is written in the following general form:

$$\frac{dQ}{dx} + \frac{dA}{dt} = -q \tag{1.3}$$

where Q = lateral discharge; and x and t = space and time coordinates. For a steady flow condition ($dA/dt = 0$) in laterals, the continuity equation, Eq. (1.3), may be rewritten as follows:

$$q + \delta q = 0; q + \frac{\delta Q}{dx} = 0; q + \frac{(Adv)}{dx} = 0\, A = \frac{dv}{dx} = -q \tag{1.4}$$

where v = velocity of flow in the lateral pipe.

Figure 1.1 illustrates the related hydraulic variables spatially varied along the horizontal lateral line with successive multiple outlets. As shown in Fig. 1.1, using the continuity equation, Eq. (1.4), the lateral discharge Q_{n+1}, at the pipe reach between successive outlets (n) and (n + 1) can be obtained from:

$$Q_{n+1} = Q_n - q_n \tag{1.5}$$

With the momentum effect resulting from decreasing the discharge in the downstream direction, the conservation of momentum equation is given in the following form:

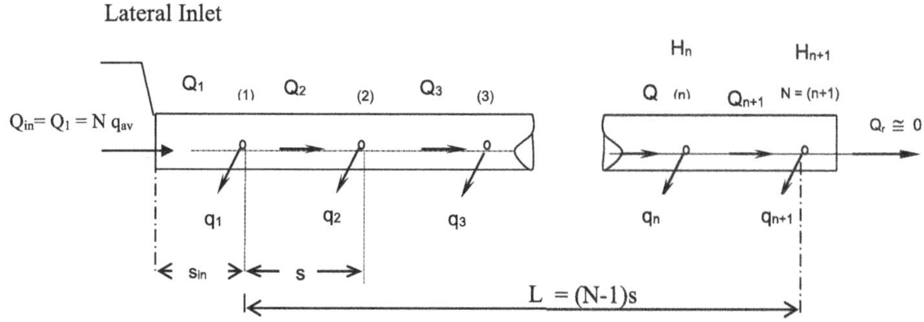

Fig. 1.1 Hydraulic Characteristics along horizontal multioutlet pipeline with successive multiple outlets

$$\Delta F = \rho(Q_{n+1} V_{n+1} - Q_n V_n) \tag{1.6}$$

where ΔF = change of pressure force; ρ = density of irrigation water; and V_n, and V_{n+1} = flow velocity between successive outlets (n − 1)~(n) and (n)~(n + 1), respectively.

Using the conservation of momentum equation, Eq. (1.6), the change of pressure head, ΔH_{n+1}, due to change of momentum between successive emitters (n) and (n + 1) may be obtained as [20, 21]:

$$\Delta H_{n+1} = \frac{\Delta p}{\gamma} = \frac{\Delta F}{A\gamma} = \frac{\rho(Q_{n+1} V_{n+1} - Q_n V_n)}{A\gamma} \tag{1.7.a}$$

$$V_{n+1} = \frac{Q_{n+1}}{A} V_n = \frac{Q_n}{A} \gamma = \rho g \tag{1.7.b}$$

$$\Delta H_{n+1} = \frac{\rho}{A\rho g}\left(\frac{Q_{n+1}^2 - Q_n^2}{A}\right) = \frac{(Q_{n+1}^2 - Q_n^2)}{gA^2} \tag{1.7.c}$$

where g = acceleration due to gravity.

Darcy-Weisbach friction formula can give accurate predictions for frictional losses based on conservation of energy principle. Therefore the following expression can be written [34–36]:

$$H_{fn+1} = f_{n+1} \frac{s}{D} \frac{Q_{n+1}^2}{2gA^2} \tag{1.8}$$

where $H_{f(n+1)}$ = friction head loss between successive emitters (n) and (n+1); f_{n+1} = Darcy-Weisbach friction coefficient for the pipe reach between successive outlets (n) and (n + 1); and D = internal diameter of the pipe.

Figure 1.2 illustrates flow variation (sudden contraction and subsequent enlargement) and related hydraulic variables along lateral section, due to presence of successive on-line emitters.

As shown in Fig. 1.2, total local energy losses, h_K' due to contraction and subsequent enlargement in upstream and downstream sections, can be evaluted from Belanger or Borda-Carnot equations:

$$h_k' = h_c + h_e = \frac{(V_c - V_r)^2}{2g} + \frac{(V_r - V_{n+1})^2}{2g} = \left(\frac{1}{C_c r} - 1\right)^2 \frac{V_{n+1}^2}{2g} = K \frac{V_{n+1}^2}{2g} \tag{1.9}$$

Finally, using the conservation of energy principle with Eqs. (1.7.c), (1.8) and (1.9) between successive outlets (n) and (n + 1), the following form can be obtained:

$$H_n + \frac{V_n^2}{2g} + z_n + h_K' = H_{n+1} + \frac{V_{n+1}^2}{2g} + z_{n+1} + H_{fn+1} + \Delta H_{n+1} + h_{kn+1}' \tag{1.10}$$

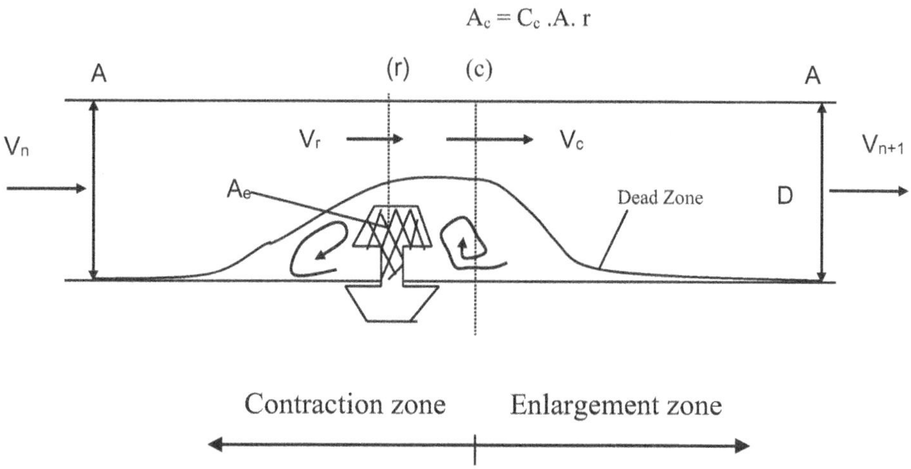

Fig. 1.2 Flow variation in two different flow zones (contraction and enlargement) and related hydraulic variables due to successive on-line emitter connections

where H_n and H_{n+1} = pressure heads for the successive outlets (n) and (n + 1); $V_n^2/2g$ and $V_{n+1}^2/2g$ = velocity heads under consideration; and z_n and z_{n+1} = elevation of successive outlets (n) and (n + 1), above an arbitrary datum.

As can be seen from the preceeding hydraulic analysis, there are five basic equations in the multioutlet pipe hydraulics. These equations are (1): the outlet discharge-pressure head relationship, Eq. (1.1); (2): the continuity equation, Eq. (1.5); (3): the Darcy-Weisbach friction formula, Eq. (1.8); (4): the Belanger or Borda-Carnot equations, Eqs. (1.9); and (1.5): the conservation of energy equation, Eq. (1.10), coupled with the conservation of momentum equation, Eq. (1.7.c). Also there are four unknown hydraulic variables (Q_{n+1}, q_{n+1}, H_n, H_{fn+1}) at any location of pipeline (n + 1), with the known previous values (Q_n, q_n ve H_n) and the other parameters (z_n, z_{n+1}, f_{n+1}, D, s, c, x, N). These unknown variables can be estimated using one of the calculation methods discussed in the following chapters.

Energy Profiles Along Multioutlet Submain Lines

Energy (EGL) and Hydraulic (HGL) Grade Lines

The concepts of energy (EGL) and hydraulic (piezometric) (HGL) grade lines are useful in hydraulic analyzing of steady pipe flow problems. The EGL in a multiple outlets pipeline will not be a straight line but a curve of the exponential type Hydraulically, flow in a multioutlet submain line is considered to be steady, spatially varied flow with decreasing discharge along the direction of flow (from upstream inlet to dowmnstream closed end). With decreasing discharge along the lateral, the energy gradient line decreases [45–48].

Types of Pressure Head Profiles for Various Uniform Line Slope Situations

The most important difficulty when recognizing the type of pressure profile is the determination of energy-grade line slope (or friction slope) (S_f), which depends on the total inlet discharge (Q) and pipe diameter (D), since Q and D are unknown parameters in most usual design cases. In the following analysis, different types of pressure head profiles (***Type-I***, ***Type-II.a***, ***Type-II.b***, ***Type-II.c*** and ***Type-III***) concerning different uniform line slope situations, will be presented in Figs. 1.3 and 1.4, respectively.

When a submain line with multiple outlets is designed using a single inlet system and laid on a uniform slope (S_0), the pressure head profile along the line resulting from the total energy drop by friction, $h_{f(L)}$ [$h_{f(L)} = S_f \times L$; S_f: the slope of the energy-gradient line or the friction slope] and the total potential energy gain (or loss) by a uniform downslope (or upslope) at the end of the line, ΔS ($\Delta S = S_0 \times L$), can be classified into three general types (Type- I, Type-II and Type-III). The classification can be based on a dimensionless energy gradient ratio, K_S ($K_S = S_0/S_f$).

When the kinetic energy is considered to be small and neglected in a submain line, the shape of profile (pressure variation) will be simply a linear combination of the operating inlet pressure head (H_{0I}), total energy drop due to friction, and total potential energy change (gain or loss) due to uniform line slope [12–14].

A comprehensive description of the energy relations for all types of pressure profiles will be presented. There are five typical pressure head profiles (Type-I, Type-II.a, Type-II.b, Type-II.c and Type-III) under three general types, and can be expressed in what follows:

Pressure Profile Type I

(Minimum Pressure at the Downstream Closed End of the Line):
The pressure head decreases with respect to the submain length. This type occurs when the submain line is laid on zero or uphill slopes. In this condition, the dimensionless energy-gradient ratio is $K_S = S_0/S_f \leq 0$. The total energy is lost by both elevation change due to uniform upslope and by friction. The maximum pressure head, H_{max} is at the inlet of the line and is equal to the operating inlet pressure, H_{0I} ($H_{0I} = H_{max}$); the minimum pressure head, H_{min}, is at the closed end of the line, and is equal to the downstream pressure head ($H_d = H_{min}$).

Pressure Profile Type II (Minimum Pressure Along the Line):
The pressure head decreases from the upstream end with respect to the pipeline length, reaches a minimum point (i_{min}), and then increases with respect to the submain line. The minimum pressure head is located somewhere along the line ($0 < i_{min} = l_{min}/L < 1$), depending on the large interval of K_S, $0 < K_S = S_0/S_f < 2.75$.

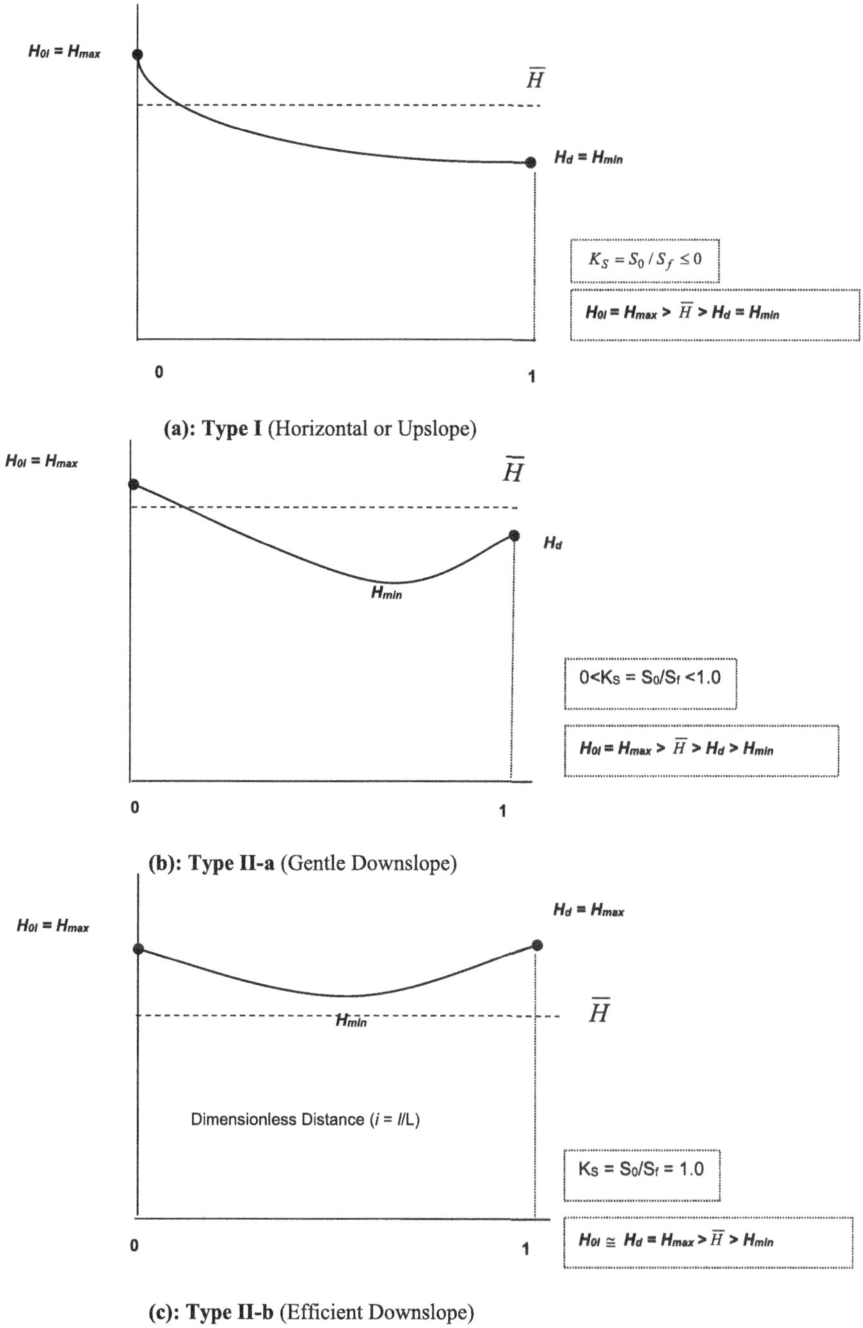

Fig. 1.3 Pressure head profiles alongs submain line for **a** Type I, **b** Type II. **a** and **c** Type II. **b** profiles

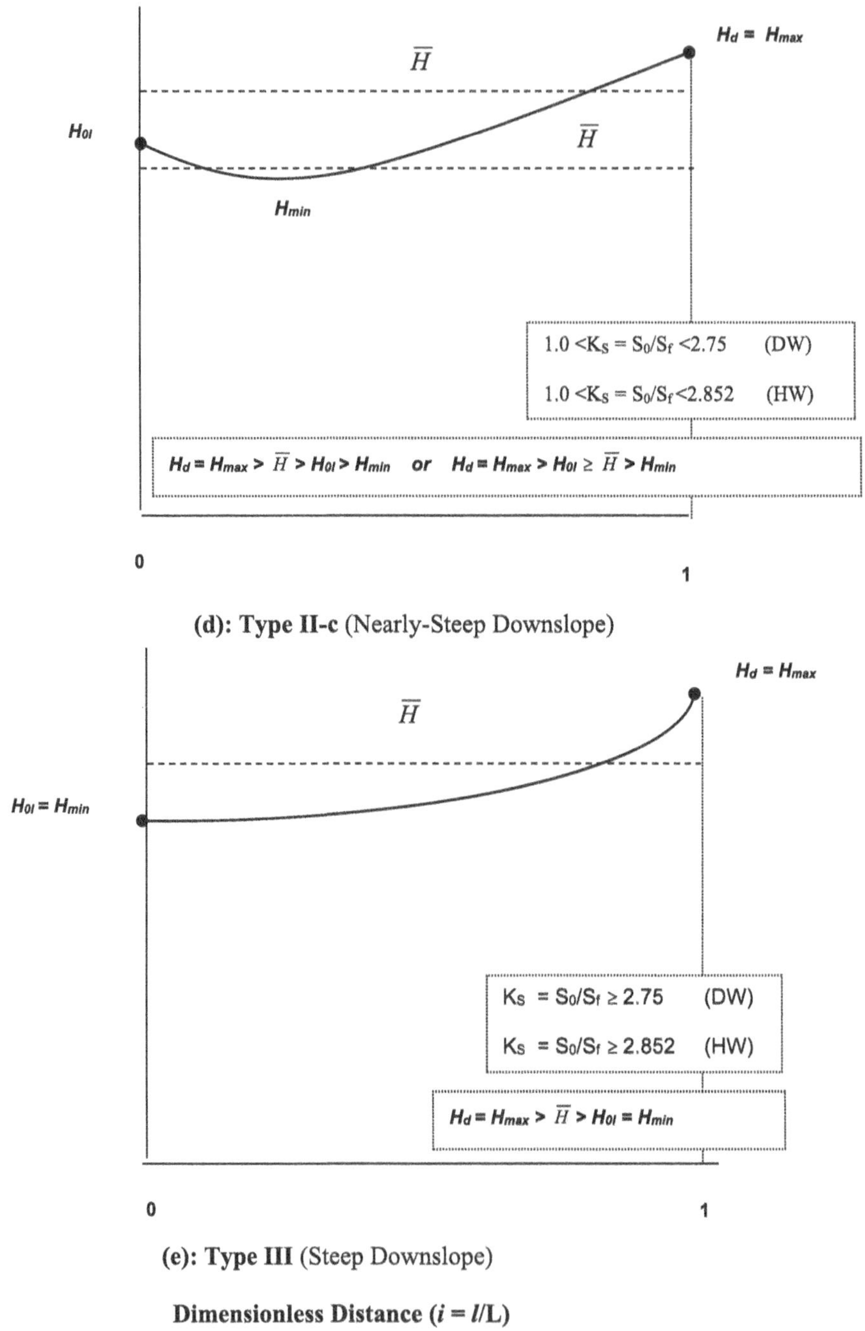

Fig. 1.4 Pressure head profiles in downslope situation for **d** Type II. **c** and **e** Type III profiles

Profile II-a:

This type of pressure profile occurs under the friction and slope situation, where $0<K_S = S_0/S_f < 1.0$. In this type, the total energy gain by uniform downslope at the end of the line is smaller than the total energy drop by friction; so the downstream end pressure head (H_d) is still less than the operating inlet pressure (H_{0I}). The maximum pressure head is at the inlet ($H_{max} = H_{0I}$); and the minimum pressure head is located somewhere along the line.

Profile II-b (Optimal Pressure Profile):

This is under the friction slope situation where the dimensionless energy-gradient ratio is $K_S = S_0/S_f = 1.0$. This profile is similar to Profile II-a, but the profile is such that the downstream closed end pressure head is equal to the operating inlet pressure head ($H_{0I} = H_d$). The maximum pressure is at the inlet ($H_{max} = H_{0I}$) as well as at the closed end of the line ($H_{max} = H_d$). The minimum pressure is located somewhere near the middle section of the line.

Profile II-c:

This is under the friction and slope situation where the dimensionless energy-gradient ratio is $1.0 < K_S < 2.75$ for the Darcy-Weisbach (DW) or $1.0 < K_S < 2.852$ for the Hazen-Williams (HW) equation. This occurs when the line slope is even steeper, so the pressure at the end of line is higher than the operating inlet pressure. In this condition, the maximum pressure is at the downstream closed end of the line ($H_{max} = H_d$), and the minimum pressure is located somewhere along the upstream segment of the line.

Profile III (Minimum Pressure at the Upstream Inlet of the Line):

This occurs when the submain line is on a steep downslope where the total energy gain by the uniform downslope is larger than the total energy drop by friction for all sections along the line. The pressure head increases with respect to the line length. This is caused by a steep downslope situation where the dimensionless energy-gradient ratio is $K_S \geq 2.75$ for the DW equation, or $K_S \geq 2.852$ for the HW equation.

In this condition the maximum pressure is at the downstream closed end of the line ($H_{max} = H_d$), and the minimum pressure head is at the pipe inlet and equal to the operating pressure head ($H_{min} = H_{0I}$).

Among all types of pressure profiles, the Type-II profile is considered as the optimal (or ideal) pressure profile which can produce the minimum pressure head difference for a given pipe length as discussed earlier [15–17]. This profile occurs when the total energy loss by friction is just balanced by the total energy gain due to uniform downslope (Profile II-b) as presented in the papers [10, 11].

Concluding Remarks

A multiple outlets pipeline is a hydraulic structure whose design is limited by the operating inlet pressure head, and by water application uniformity that is affected by the total energy loss along the pipeline, the field topography, as well as the outlet hydraulic characteristics. This chapter presents the fundamental concepts and principles upon which the hydraulics of multioutlets pipeline systems is based. The review is intended to be sufficiently complete that readers who have taken a good first course in elementary multioutlet pipe flow hydraulics, various design concepts with engineering applications in profession, covering different types of pipe design problems that are the foundation of the material in this book. An introduction to the fundamental equations that are the foundation of most of the subsequent developments in the book, is presented. Because the concept of the Energy Grade Line (EGL) and the Hydraulic Grade Line (HGL) is so useful, we shall look at this idea separately. The chapter concludes the governing equations for steady-state pipe flow that will allow readers to better understand the later analytical derivations for the coming chapters.

Acknowledgements The senior author wants to express his acknowledges to Prof. Vijay. P. Singh at Texas A&M University, Emeritus Prof. Necati Ağıralioğlu from International Antalya Science University, and Distinguished Emeritus Professor Zekai Şen from Istanbul Medipol University, for their highly encouragement on preparing this book, many scientific activities, projects and so on. The Scientific and Technological Research Council of Turkey (TUBITAK) is acknowledged for supporting by the fellowships and grant program during the studies in the Texas A&M University/BAEN. We also thank to publishers, potential readers and reviewers for providing constructive reviews and criticism on the content of the book, in advance.

Conflict of Interest The author/editor has no conflicts of interest, financial or otherwise, to declare.

References

1. H. Anyoji, I.P. Wu, Statistical approach for drip lateral design. Trans. ASAE **30**(1), 187–192 (1987)
2. V.F. Bralts, I.P. Wu, H.M. Gitlin, Manufacturing variation and drip irrigation uniformity. Trans. ASAE **24**(1), 1234–1240 (1981)
3. V.F. Bralts, D.M. Edwards, I.P. Wu, Drip irrigation design and evaluation based on the statistical uniformity concept. Adv. Irrig. **4**, 67–117 (1987) (Academic, New York)
4. J.E. Christiansen, Irrigation by sprinkling. Calif. Agric. Exp. Stn. Bull. **670**, (1942) (Univ. of California, Davis, California)
5. R.H. Cuenca, *Irrigation systems design: an engineering approach* (Prentice Hall, New Jersey, 1989)
6. J. Keller, R.D. Bliesner, *Sprinkler and trickle irrigation* (Van Nostrand Reinhold, New York, 1990)

7. G. Yıldırım, N. Ağıralioğlu, Variation of total friction head losses and uniformity coefficients based on inlet pressure head in trickle laterals, in *Proceedings of* the *International Conference on Water Resources Management in Arid Regions* (Kuwait Institute for Scientific Research, Kuwait, 2002)
8. G. Yıldırım, N. Ağıralioğlu, Comparative analysis of hydraulic calculation methods in design of microirrigation laterals. J. Irrig. Drain. Eng. ASCE **130**(3), 201–217 (2004)
9. G. Yıldırım, N. Ağıralioğlu, Variation of design parameters in microirrigation laterals. ARI Interdiscip. J. Phys. Eng. Sci. **54**(1), 42–53 (2004)
10. G. Yildirim, V.P. Singh, Operating pressure for uniformly sloping submain lines with different types of pressure profiles. I: Improved energy-gradient-ratio (EGR) approach. 6th International Perspective on Water Resources and the Environment (ASCE&EWRI, Izmir, Turkey, 2013)
11. G. Yildirim, V.P. Singh, Operating pressure for uniformly sloping submain lines with different types of pressure profiles. II: Application and Verification. 6th International Perspective on Water Resources and the Environment (ASCE&EWRI, Izmir, Turkey, 2013)
12. I.P. Wu, H.M. Gitlin, Hydraulics and uniformity for drip irrigation. J. Irrig. Drain. Div. ASCE, **99**(IR2), 157–168 (1973) (Proc. Paper 9786)
13. I.P. Wu, H.M. Gitlin, Drip irrigation design based on uniformity. Trans. ASAE **17**(3), 429–432 (1974)
14. I.P. Wu, H.M. Gitlin, Energy gradient line for drip irrigation laterals. J. Irrig. Drain. Eng. ASCE **101**(4), 323–326 (1975)
15. I.P. Wu, Energy gradient line approach for direct hydraulic calculation in drip irrigation design. Irrig. Sci. **13**, 21–29 (1992)
16. I.P. Wu, An assessment of hydraulic design of microirrigation systems. Agric. Water Manage. **32**, 275–284 (1997)
17. I.P. Wu, R. Yue, Drip lateral design using energy gradient line approach. Trans. ASAE **36**(2), 389–394 (1993)
18. G. Yıldırım, Comparative analysis of hydraulic calculation methods in design oh micro-irrigation laterals and linear solution of hydraulic analysis of tapered laterals. PhD. thesis, Istanbul Technical University, Hydraulics Division, Istanbul, Turkey, 2007
19. G. Yıldırım, N. Ağıralioğlu, Linear solution for hydraulic analysis of tapered microirrigation laterals. J. Irrig. Drain. Eng. ASCE 130(1), 78–87 (2004)
20. H.M. Hathoot, A.I. Al-Amoud, F.S. Mohammad, Analysis and design of trickle irrigation laterals. J. Irrig. Drain. Eng. ASCE **119**(5), 756–767 (1993)
21. H.M. Hathoot, A.I. Al-Amoud, F.S. Mohammad, Analysis and design of sprinkler irrigation laterals. J. Irrig. Drain. Eng. ASCE **120**(3), 534–549 (1994)
22. G. Yıldırım, Computer-aided design of micro-irrigation laterals by using LatCAD. MSc. thesis, Istanbul Technical University, Hydraulics Division, Istanbul, Turkey, 2001
23. Y. Kang, S. Nishiyama, Analysis and design of microirrigation laterals. J. Irrig. and Drain. Eng. ASCE **122**(2), 75–82 (1996)
24. Y. Kang, S. Nishiyama, Design of microirrigation submain unit. J. Irrig. Drain. Eng. ASCE **122**(2), 83–89 (1996)
25. E.J. Scaloppi, R.G. Allen, Hydraulics of irrigation laterals: comparative analysis. J. Irrig. Drain. Eng. ASCE **119**(1), 91–115 (1993)
26. A.W. Warrick, M. Yitayew, Trickle lateral hydraulics.I: analytical solution. J. Irrig. Drain. Eng. ASCE **114**(2), 281–288 (1988)
27. G.Z. Watters, J. Keller, Trickle irrigation tubing hydraulics, in *presented at the Summer Meeting of ASAE* (Utah State University, Logan, Utah, Paper No. 78-2015, 1978)

28. G. Yıldırım, O. Kose, New software (Multi-flowCAD) development for sub-main line hydraulic design: computational algorithm, computer visualization and implementation. Arab. J. Sci. Eng. **41**, 1551–1560 (2016)
29. G.Yıldırım, Computer-Based Analysis of Hydraulic Design Variables for Uniformly Sloping Microirrigation System Laterals. J. Irrig. Drain. Eng. ASCE **141**(7), 04014079 (2015)
30. G. Yıldırım, An Assessment for Hydraulically Efficient Design of Uniformly Sloping Submain, Lines. Irrig. Drain. **64**(2), 157–168 (2015)
31. G. Yıldırım, Total energy loss assessment for trickle lateral lines equipped with integrated in-line and on-line emitters. Irrig. Sci. **28**(4), 341–352 (2010)
32. G. Yıldırım, V.P. Singh, A *MathCAD* procedure for commercial pipeline hydraulic design considering local energy losses. Adv. Eng. Softw. **41**(2), 489–496 (2010)
33. G. Yıldırım, Computer-based analysis of explicit approximations to the implicit Colebrook-white equation in turbulent flow friction factor calculation. Adv. Eng. Softw. **40**(11), 1183–1190 (2009)
34. G. Yıldırım, M. Özger, Neuro-fuzzy approach in estimating Hazen-Williams friction coefficient for small-diameter polyethylene pipes. Adv. Eng. Softw. **40**(8), 593–599 (2009)
35. G. Yıldırım, Simplified procedure for hydraulic design of small-diameter plastic pipes. Irrig. Drain. (ICID J.) **58**(3), 209–233 (2009)
36. M. Özger, G. Yıldırım, Determining turbulent flow friction coefficient using adaptive neuro-fuzzy computing technique. Adv. Eng. Softw. **40**(4), 281–287 (2009)
37. G. Yıldırım, determining operating ınlet pressure head ıncorporating uniformity parameters for multioutlet plastic pipelines. J. Irrig. Drain. Eng. ASCE **134**(3), 341–348 (2008)
38. G. Yıldırım, An assessment of hydraulic design of trickle laterals considering effect of minor losses. Irrig. Drain. (ICID J.) **56**(4), 399–421 (2007)
39. G. Yıldırım, Analytical relationships for designing multiple outlets pipelines. J. Irrig. Drain. Eng., ASCE **133**(2), 140–154 (2007)
40. G. Yıldırım, Hydraulic analysis and direct design of multiple outlets pipelines laid on flat and sloping lands. J. Irrig. Drain. Eng. ASCE **132**(6), 537–552 (2006)
41. G. Yıldırım, Discussion of "Statistical analysis of non-pressure-compensating and pressure-compensating drip emitters" by H. Perea, J. Enciso-Medina, V. P. Singh, D. P. Dutta, and B. J. Lesikar. J. Irrig. Drain. Eng. ASCE (2016). https://doi.org/10.1061/(ASCE)IR.1943-4774.0000644
42. S.W. Churchill, Friction factor equation spans all fluid-flow regimes. Chem. Eng. **84**(24), 91–92 (1977)
43. R.D. von Bernuth, Simple and accurate friction loss equation for plastic pipe. J. Irrig. Drain. Eng. ASCE **116**(2), 294–298 (1990)
44. R.D. von Bernuth, T. Wilson, Friction factors for small diameter plastic pipes. J. Hydraulic Eng. ASCE **115**(2), 183–192 (1989)
45. G. Yıldırım, N. Ağıralioğlu, Discussion of 'Continuous outflow variation along irrigation laterals: effect of the number of outlets' by J. D. Valiantzas. J. Irrig. Drain. Eng. ASCE **129**(5), 382–386 (2003)
46. G. Yıldırım, N. Ağıralioğlu, Discussion of 'Inlet pressure, energy cost, and economic design of tapered irrigation submains' by J. D. Valiantzas. J. Irrig. Drain. Eng. ASCE **131**(2), 220–224 (2005)
47. G. Yıldırım, N. Ağıralioğlu, Closure to 'Linear solution for hydraulic analysis of tapered microirrigation laterals' by G. Yıldırım and N. Ağıralioğlu. J. Irrig. Drain. Eng. ASCE **131**(5), 490–491 (2005)
48. M. Yitayew, A.W. Warrick, Trickle lateral hydraulics.II: design and examples. J. Irrig. Drain. Eng. ASCE **114**(2), 289–300 (1988)

2 New Analytical Development: Revised-Energy-Gradient Line (REGL) Method Based on Average Friction Drop Ratio Approach: Methodology, Implementation and Design Applications

Introduction

Pressurized pipelines with multiple outlets (laterals and manifolds) are extensively used to distribute water in irrigated areas under different types of irrigation systems such as surface, sprinkler, and trickle. Adequate analysis of hydraulic design of multiple outlets pipeline is important for the proper performance of these systems. Multiple outlet pipeline is a hydraulic structure whose design is limited by the pressure head at the pipeline inlet, water application uniformity and, consequently by the total head losses due to friction. In previous studies [1–18] for hydraulic computation of multiple outlets pipelines, the primary solution is based on an outflow that is uniform, based on the derived hydraulic profile. However, this basic assumption may cause significant errors in hydraulic computations.

Theoretical Background: Uniform and Non-uniform Outflow Profile Concepts

In the previous traditional approaches [1–18] for hydraulic computation of multiple outlets pipelines, the primary solution is based on an outflow that is uniform, although ramifications of the manufacturer's variability have been modeled based on the derived hydraulic profile. However, this basic assumption may cause significant errors in hydraulic computations.

The energy-gradient line (EGL) method is traditionally used for the determination of the lateral pressure head and outflow profiles [15–17]. This technique is simple; however, it is subject to certain errors since the energy gradient line is determined by the assumption of constant emitter outflow [13].

G. Yıldırım, *Hydraulic Principles and Design Concepts for Submain Units with Multiple Outlet Pipelines*, Synthesis Lectures on Engineering, Science, and Technology,
https://doi.org/10.1007/978-3-031-32495-6_2

Two types of errors are caused by this simple EGL approach; one is associated with the shape of energy gradient line and the other is the total friction drop at the end of the line. Based on this analysis of possible errors, for turbulent flow emitter and an emitter flow variation equal or less than 10% the simple EGL approach is used for drip irrigation design. For non-turbulent flow emitters or when the emitter flow variation is larger than 10%; designs can be made by using an adjusted total discharge or revised energy gradient line (REGL) approach [13, 14, 18]. The application of the analytical EGL (or REGL) method is very easy; however, it has a limitation of application region due to constraints in emitter flow regimes and emitter flow variation along the lateral [19].

The authors [4] derived a differential approach to solve hydraulic problems of various types of multiple outlet pipes with uniform and nonuniform outflow, which simulates pressure head distributions along uniform sprinkler and trickle laterals and manifolds. An alternative treatment in trickle laterals, authors presented [11, 16] a non-linear second-order, ordinary differential equation which includes a spatially variable discharge function. It dismissed the assumption of a uniform emitter flow along the lateral, as suggested in methods presented earlier [1, 17].

The authors [20, 21] presented a stepwise calculation method for analysis and design of trickle and sprinkler laterals. The step-by-step (SBS) method has the highest accuracy in reference to the recent studies [7, 8, 22, 23] because only the basic equations of the hydraulics of steady pipe flow were used in each part of the lateral, which was divided by the individual consecutive outlets. In the last decade, the finite-element method (FEM) has been applied to analyze hydraulics of single and paired microirrigation laterals [22] and submain units [23] on flat and sloped fields. The method has been found to be very accurate for designing microirrigation laterals.

The authors [7, 8] proposed the successive-approximations method (SAM) for solving lateral hydraulic problems for laminar or turbulent flow. However, the SAM method can be suitable for didactic applications since the solution requires large amounts of the execution time and calculation effort. The author [5, 6] derived an analytical energy line of a single-diameter lateral with taking into account the effect of the number of outlets. However, several relationships presented in these studies were discussed in the recent papers [24–30] based on the noteworthy opinion [31], related to the basic derivative procedure in the background of some analytical expressions.

In a recent study, the state-of-the art review of some recent studies for microirrigation lateral design was presented [27]. In the analysis, a detailed comparison test was applied to several design examples to cover various combinations of design parameters and varying emitter discharge exponents, then the dimensionless design charts for different ranges of lateral diameter and length and for various uniform line slope cases, are also presented.

Determination of Energy-Gradient Line (EGL) and Hydraulic Grade Line (HGL)

The concepts of energy (EGL) and hydraulic (piezometric) (HGL) grade lines are useful in hydraulic analyzing of steady pipe flow problems. The EGL or HGL in a multiple outlets pipeline will not be a straight line but a curve of the exponential type. Figure 2.1 shows the sketch of the horizontal pipe with multiple outlets, and the shapes of the hydraulic and energy gradient curves along its length. Flow in a multiple outlets pipeline is considered to be steady, spatially varied flow with decreasing discharge along the direction of flow [32–44].

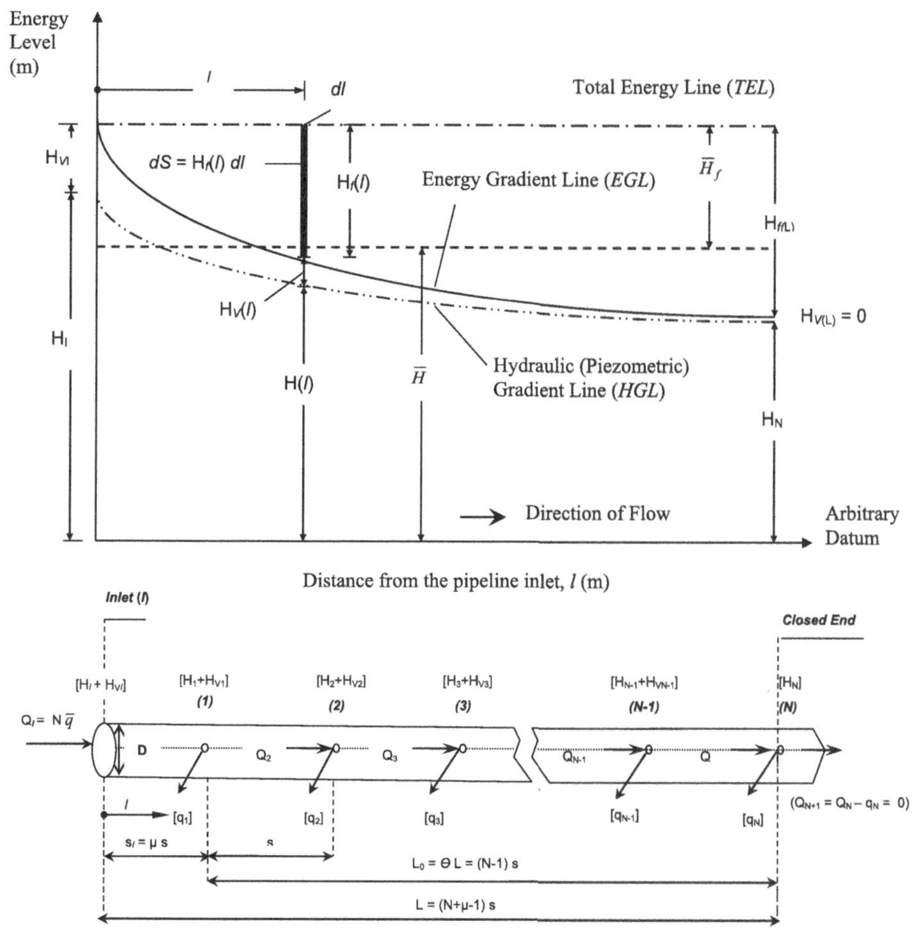

Fig. 2.1 Scheme of horizontal pipe with successive multiple outlets, with the shapes of hydraulic (piezometric) and energy-grade lines along its length

Mathematical Formulation

Discharge and Velocity Head

Typically, the discharge through each outlet is governed by the following relationship:

$$q = kH^x \tag{2.1}$$

in which q = discharge through the outlet; H = operating pressure head of the outlet; k and x = coefficient and exponent of the outlet discharge, respectively. Assuming that outlets along the pipeline have uniform outflow $\overline{q}$, then the total flow rate that passes through the pipe section located at any distance l from the pipeline inlet Q(l) is commonly given as [4, 17]

$$Q(l) = \left(N - \frac{l}{s}\right)\overline{q} = Q_I(1 - i) \tag{2.2}$$

where Q(l) = the discharge at a section of length l measured from the pipeline inlet; $Q_I = N\overline{q}$ = total inflow rate for the entire length of the pipeline; N = total number of equally spaced operating outlets for the entire pipeline; s = outlet spacing; and i = dimensionless variable of distance i = l/L, which is the percentage of the length to the total length.

According to Eq. (2.2) the average flow velocity also varies linearly along the pipeline and, the velocity head gradient can be written as follows

$$S_V(l) = \frac{2H_{VI}}{L}(i - 1) \tag{2.3}$$

where $S_V(l)$ = velocity head gradient at any distance l; and $H_{VI} = 8(N\overline{q})^2/\pi^2 gD^4$ = velocity head at the inlet of the pipe; and D and L = internal diameter and length of the pipe.

Integrating the gradient of the velocity head given by Eq. (2.3) between the limits 0 and l and then, the velocity head at any distance l is obtained as

$$H_V(l) = H_{VI}(1 - i)^2 \tag{2.4}$$

where $H_V(l)$ = velocity head for the pipe section located at any distance l.

Flow Regimes

Flow in a submain line is usually turbulent (3,000 < R < 10^5; R: Reynolds number). Sometimes fully turbulent flow (10^5 < R < 10^7) exists at the upstream end, and flow becomes laminar (R < 2,000) at the downstream reach where the flow velocity decreases to zero. The transition flow approximately occurs in the interval 2,000 < R ≤ 3,000, and the flow characteristics are indeterminate for the transition zone. Further information can be available from the literature [2].

Friction Head Loss

The general form of the Darcy-Weisbach friction formula which proved to be the most accurate and reliable [9, 10, 12] can be approximated by the following expression:

$$S_f(l) = \frac{dH_f(l)}{dl} = \frac{f}{D}H_V(l) = \frac{C}{D^n}Q^m(l) \tag{2.5}$$

$$C = \frac{8c_1(\nu/4)^{2-m}}{g\pi^m} \tag{2.6}$$

In the rest where $S_f(l)$ = friction slope or the slope of the energy grade line at any distance l from the pipe inlet expressed as the energy loss per unit length of the pipeline, and per unit weight of fluid; $H_f(l)$ = friction head loss per unit weight of fluid at any distance l; f and C = friction coefficient and combined friction coefficient for the Darcy-Weisbach formula assuming to be constant along the pipe; c_1 = constant for the particular friction coefficient formula used; m and n = velocity and diameter exponents, respectively. Generally n is approximately equal to m+3 [45–48]; ν = kinematic viscosity of water; and g = acceleration due to gravity.

Note that, the combined friction coefficient C is appropriate for plastic pipes (PVC or PE). For aluminum pipes, the friction coefficient f is given by the well-known Churchill equation [49, 50]

$$f = 8\left[(8/R)^{12} + 1/(A+B)^{3/2}\right]^{1/12} \tag{2.7}$$

where A and B = empirical parameters for computing Darcy-Weisbach friction coefficient. A and B are calculated by

$$A = \left\{2.457\ln\left[1/\left((7/R)^{0.9} + 0.27(\varepsilon/D)\right)\right]\right\}^{16};\quad B = (37,530/R)^{16} \tag{2.8}$$

where ε = equivalent roughness height of the internal pipe surface.

Integrating Eq. (2.5) with Eq. (2.2) between the limits 0 and l yields [17, 45]

$$H_f(l) = H_{f(L)}\left[1 - (1-i)^{m+1}\right] \tag{2.9}$$

$$H_{f(L)} = FH_{f0} = \frac{1}{(m+1)}H_{f0} \tag{2.10}$$

$$H_{f0} = f\frac{L}{D}H_{VI} = C\frac{Q_I^m}{D^n}L \tag{2.11}$$

where $H_{f(L)}$ = total friction drop at the closed end of the pipeline; and H_{f0} = total friction drop in a similar pipeline without outlets conveying the total flow rate along its entire length. It should be noted, the second multiplying term in the bracket in the right-hand side of Eq. (2.9) can be described as *the friction drop ratio*, $R_i = H_f(l)/H_{f(L)}$.

An alternate version for the total friction drop $H_{f(L)}$ may be obtained by replacing in Eq. (2.10) the traditional correction factor F by the Christiansen's correction factor F_C [16]

$$H_{f(L)} = F_C H_{f0} \tag{2.12}$$

Friction Coefficient (*f*)

The general form of the friction coefficient (f) in terms of the Reynolds number, is given by the simple expression:

$$f = c_1[R_I]^{m-2} \tag{2.13}$$

$R_I = (4N\overline{q}/\pi D\nu)$ = Reynolds number at the inlet of the pipe.

For laminar flow regime (R < 2,000; R = Reynolds number for pipe flow), m = 1.0, n = 4.0, c_1 = 64 and C = 4.195 × 10^{-6}; for turbulent flow regime in a smooth pipe (3,000 < R ≤ 10^5), m = 1.75, n = 4.75, c_1 = 0.316 (Blasius equation) and C = 7.792 × 10^{-4}, and finally for fully turbulent flow regime (10^5 < R < 10^7), m = 1.828, n = 4.828, c_1 = 0.130 [12] and C = 9.589 × 10^{-4}, are taken.

Friction Correction Factor (*F*)

To compute total friction drop at the closed end of the multiple outlets pipeline requires calculating the friction head loss for each pipe segment between consecutive outlets, in a stepwise manner. This procedure is characterized by a stepwise correction factor, F_{SW} computed for finite number of outlets as given by:

$$F_{SW} = \frac{1}{N^{m+1}} \sum_{K=1}^{K=N} K^m \tag{2.14}$$

where K = an integer represents the consecutive order of outlets; and m = velocity (or flow rate) exponent for the particular friction head loss formula used.

For the traditional approach, the friction correction factor takes the form

$$F = \frac{1}{(m+1)}. \tag{2.15}$$

Equation (2.15) represents the friction loss correction factor that takes into account the continuous and uniform outflow along the pipeline. This correction factor is essentially equivalent to that proposed [48], provided the number of outlets is infinity.

The correction factor F_C, for the case in which the first outlet is located in a full outlet spacing from the pipeline inlet, is of the form

$$F_C = \frac{1}{(m+1)} + \frac{1}{2N} + \frac{\sqrt{m-1}}{6N^2} \tag{2.16}$$

If the first outlet is half outlet spacing from the pipeline inlet, F_C is given by [50]

$$F_C = \frac{2N}{(2N-1)}\left(\frac{1}{m+1} + \frac{\sqrt{m-1}}{6N^2}\right) \tag{2.17}$$

Comparison Test for Friction Correction Factors

Christiansen's approach [48] given by Eqs. (2.16) and (2.17) is based on uniform and discrete outflow from a finite number of operating outlets. Therefore, significant differences between the two correction factors given by Eq. (2.15) and Eqs. (2.16) and (2.17) are expected especially, when the number of outlets in the pipeline is small [4].

In order to do a systematic comparison, three correction factors (F_{SW}, F and F_C) are computed for the value of the velocity exponent m = 1.75 (Darcy-Weisbach formula and turbulent flow in a smooth pipe), full outlet spacing from the pipe inlet (s_I = s), and for various number of operating outlets (presented in the logarithmic scale), then the results were plotted in Fig. 2.2. In this figure, the secondary axis of y (dotted line) represents the percentage differences between three correction factors F_{SW}, F and F_C.

As shown in Fig. 2.2, the traditional factor F [Eq. (2.15)] yields the constant value as 0.3636 that is independent of the number of outlets N. The factor F_C [Eq. (2.16)] yields almost the same values with the stepwise correction factor F_{SW} [Eq. (2.14)] for various number of outlets.

The comparison test shows that the Christiansen's factor F_C accurately predicts the friction head loss since F_C is an accurate approximation of the stepwise correction factor F_{SW}. The traditional factor F tended to give underpredicted values systematically with respect to the factors F_C and F_{SW}, particularly, as the number of outlets decreases, the difference from these factors (F_C and F_{SW}) being more significant. For instance, for a large value of N = 100, the difference 1.4%, and for a value of N = 10, the difference 14% whereas for an extreme small value of N = 5, the difference 29%.

Average Friction Drop Ratio Approach for Energy-Gradient Line (EGL) Determination

In the following section, the improved energy-grade line of a multiple outlets pipe will be determined based on the average friction drop ratio approach with a simple exponential function, to express the nonuniform outflow concept. In the first stage of the analytical development the preliminary energy relations for the traditional uniform outflow profile [3, 47, 51–53], will be presented herein.

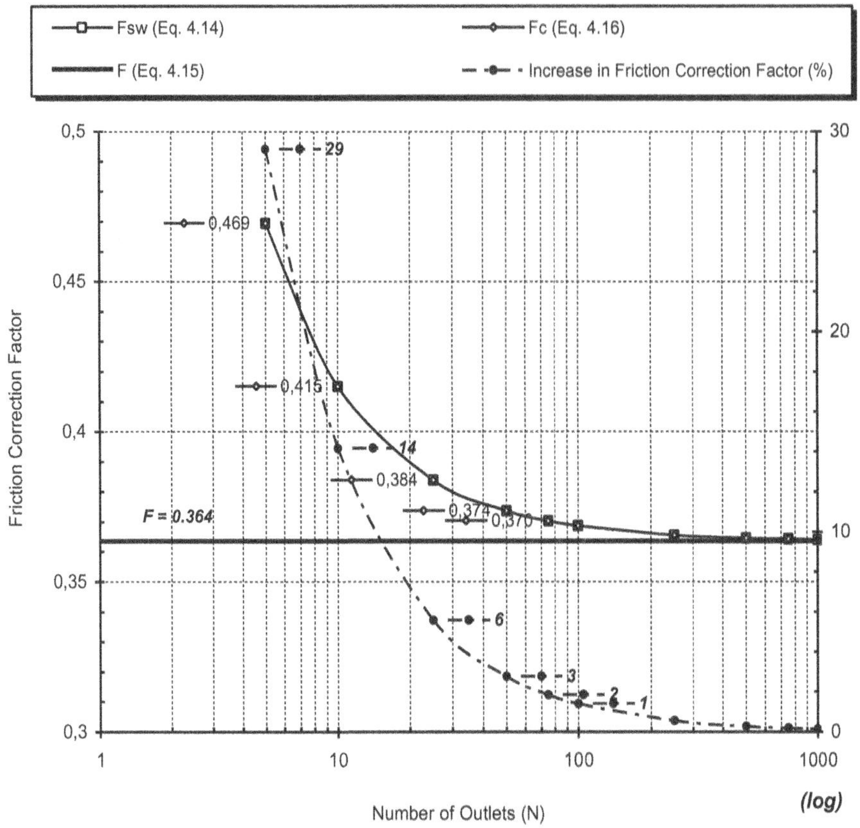

Fig. 2.2 Variation of the friction correction factors (F, F_{SW} and F_C) and percentage differences between these factors (dotted line) for the Darcy-Weisbach friction formula (m = 1.75) and full outlet spacing from the inlet, with respect to the various number of operating outlets

Pressure Head Profile Based on Uniform Outflow Concept

As shown in Fig. 2.1, let H(l) be the pressure head at a given length l from the pipe inlet, H_{0I} be the operating inlet pressure head, and S_0 be the uniform slope of the pipeline. Assuming the average pressure head for the entire pipeline is located in the middle of the length of the pipeline, and applying conservation of energy principle between the pipe segments $l = 0$ and $l = l$, then pressure head distribution along the pipeline is expressed by

$$H(l) = H_I + H_{VI}\left[1 - (1-i)^2\right] - H_{f(L)}\left[1 - (1-i)^{m+1}\right] - pi\Delta S; \quad \Delta S = S_0 L \tag{2.18}$$

in which ΔS = total head loss or gain due to elevation at the closed end of the pipeline. Note that in this analysis, the slope of a downhill pipeline is negative ($p = -1$), and the slope of an uphill pipeline is positive ($p = 1$).

Let $\overline{H}$ be the average pressure head along the pipeline can be written

$$\overline{H} = H_I - (\overline{H}_f + p\overline{\Delta S}); \qquad \overline{\Delta S} = \frac{\Delta S}{2} \tag{2.19}$$

in which $\overline{H}_f$ = average head loss due to friction along the pipeline; and $\overline{\Delta S}$ = average head loss or gain due to elevation.

To find out the average head loss due to friction along the pipeline $\overline{H}_f$, a strip of length dl is considered at any length from the pipeline inlet as shown in Fig. 2.1.

Let dS be the area of the strip, then the average head loss due to friction along the pipeline $\overline{H}_f$ is expressed as

$$\overline{H}_f = \frac{1}{L}\int_{l=0}^{l=L} dS = \frac{1}{L}\int_{l=0}^{l=L} H_f(l)dl = \left(\frac{m+1}{m+2}\right)H_{f(L)} \tag{2.20}$$

Equation (2.20) yields for m = 1.75, the average head loss due to friction along the pipeline is about 73% of the total friction drop at the downstream closed end of the pipeline, when the Darcy-Weisbach formula is used.

Substituting the expression for $\overline{H}_f$ from Eq. (2.20) into Eq. (2.19) then, subtracting from Eq. (2.18), finally one can obtain the following expression for the pressure head distribution based on uniform outflow concept:

$$H(l) = \overline{H} + H_{VI}\left[1 - (1-i)^2\right] + H_{f(L)}\left[(1-i)^{m+1} - \frac{1}{(m+2)}\right] - p\Delta S\left(i - \frac{1}{2}\right) \tag{2.21}$$

Nonuniform Outflow Profile Concept Based Exponential Function

To this point the preliminary analytical derivations presented above are similar to the referred paper [1]. In the following analytical derivation, the energy-grade line will be improved for the nonuniform outflow profile.

Instead of the traditional Eq. (2.2), to describe variation of flow rate along the pipeline Q(l), a simple but sufficiently accurate exponential function [16–18, 33] is given by

$$Q(l) = Q_I(1-i)^{\beta} \tag{2.22}$$

where β = nonuniform outflow exponent; then the outflow from the each multiple outlet will not be uniform along the pipeline.

If Eq. (2.22) is evaluated for the upper half segment of the pipeline ($i = 0.5$), then arranged with the expression for the flow rate entering over the upper half segment of the pipeline, $Q_{0.5} = (kN/2)\overline{H}_{0.5}^{x}$, finally the β can be evaluated from

$$\beta = 1 - 3.322x \log\left(\frac{\overline{H}_{0.5}}{\overline{H}}\right) \tag{2.23}$$

where $\overline{H}_{0.5}$ = average pressure head along the upper half segment of the pipeline.

Assuming the average pressure head for the upper half segment of the pipeline $\overline{H}_{0.5}$ is located in the middle of the length of the upper half segment of the pipeline, and applying the conservation of energy principle between the pipe segments l = L/4 and l = L, then the $\overline{H}_{0.5}$ can be expressed as

$$\overline{H}_{0.5} = H_N + H_{f(L)} + (p\overline{\Delta S}_{0.5} - \overline{H}_{f0.5}); \quad \overline{\Delta S}_{0.5} = \frac{3}{4}\Delta S \tag{2.24}$$

where H_N = downstream pressure head at the first outlet from the closed end of the pipeline; $\overline{H}_{f0.5}$ = average head loss due to friction along the upper half segment of the pipeline; and $\overline{\Delta S}_{0.5}$ = average head loss or gain due to elevation at the end of the upper-quarter segment of the pipeline.

Downstream closed end pressure head H_N can be obtained for i = 1 using the expression of pressure head distribution given by Eq. (2.21) as follows

$$H_N = \overline{H} + H_{VI} - \left(\frac{1}{m+2}\right)H_{f(L)} - \frac{p\Delta S}{2} \tag{2.25}$$

Average head loss due to friction between the pipe segments l = 0 and l = L/2 for the upper half segment of the pipeline $\overline{H}_{f0.5}$ can be expressed as

$$\overline{H}_{f0.5} = \frac{2}{L}\int_{l=0}^{l=L/2} dS = \frac{2}{L}\int_{l=0}^{l=L/2} H_f(l)dl = \left(\frac{m + 0.5^{m+1}}{m+2}\right)H_{f(L)} \tag{2.26}$$

Equation (2.26) yields for m = 1.75, the average head loss due to friction along the upper half segment of the pipeline is about 51% of the total friction drop at the downstream closed end of the pipeline, when the Darcy-Weisbach formula is used.

Combining the expression for $\overline{\Delta S}_{0.5}$ from Eq. (4.24) with Eq. (2.25) and Eq. (2.26) into Eq. (2.24), and dividing by $\overline{H}$ then, substituting into Eq. (2.23), finally the β can be computed from:

$$\beta = 1 - 3.322x \log\left[1 + \frac{H_{VI}}{\overline{H}} + \lambda\frac{H_{f(L)}}{\overline{H}} + \frac{1}{4\overline{H}}(p\Delta S)\right]; \quad \lambda = \left(\frac{1 - 0.5^{m+1}}{m+2}\right) \tag{2.27}$$

From this point of the analysis, the energy grade line for nonuniform outflow distribution can be determined by the following Eq. (2.27) instead of Eq. (2.21), in which the velocity exponent m should be replaced by its improved value as $m_\beta = \beta \times m$:

$$H(l) = \overline{H} + H_{VI}\left[1 - (1-i)^2\right] + H_{f(L)}\left[(1-i)^{m_\beta+1} - \frac{1}{(m_\beta+2)}\right] - p\Delta S\left(i - \frac{1}{2}\right) \tag{2.28}$$

Operating Inlet Pressure Head Determination

Inlet pressure head H_{0I} can be evaluated from Eq. (2.28) for the value of variable of distance, i = 0, as follows:

$$H_{0I} = \overline{H} + \frac{(m+1)}{(m+2)} H_{f(L)} + p\frac{\Delta S}{2} \tag{2.29}$$

Extreme Pressure Heads with Extreme Outflow Positions

Consider a pipeline with all multiple outlets being equally spaced then, the inlet spacing s_I, from the pipeline inlet to the first outlet is expressed as the some fraction, μ ($0 \le \mu \le 1.0$) of the common outlet spacing s, $s_I = \mu s$, and it usually taken as 0.5 or 1.0 for multiple outlets pipelines.

Substituting the expression for i_{max} ($i_{max} = \mu$ s/L) into Eq. (2.28), then the maximum pressure head H_{max} for uphill and zero slope, can be obtained as

$$H_{\max} = \overline{H} + H_{VI}(1-\theta^2) + H_{f(L)}\left[\theta^{m_\beta+1} - \frac{1}{(m_\beta+2)}\right] + p\Delta S\left(\theta - \frac{1}{2}\right) \tag{2.30}$$

where $\theta = L_0/L$, length ratio which is the percentage of the distance between the first and last outlets to the total length of the pipeline; $L_0 = (N - 1)$ s = length of the distance between the first and last outlets except for the inlet spacing, s_I; $L = (N + \mu - 1)$ s = total length of the pipeline.

Similarly, the minimum pressure head H_{min} for uphill and zero slope cases, is obtained from Eq. (2.28) for $i_{min} = 1.0$, which is essentially equivalent to the downstream end pressure head H_N given by Eq. (2.25).

For downhill slope case, the location of the maximum pressure head is found either to the first outlet near the pipeline inlet ($i_{max} = \mu$ s/L) or at the first outlet from the downstream closed end point ($i_{min} = 1.0$).

The location of the minimum pressure head for downhill line is determined where the first derivative of the pressure head distribution given by Eq. (2.28) is equal to zero. For simplification of this derivative procedure, the velocity head term can be omitted in Eq. (2.28), then the location i_{min} where the minimum pressure head occurs, can be approximated by

$$i_{min} = \frac{l_{min}}{L} = 1 - \left[\frac{(-p\Delta S)}{(m_\beta + 1)H_{f(L)}}\right]^{1/m_\beta} \tag{2.31}$$

Substituting Eq. (2.31) for i_{min} into Eq. (2.28), then the H_{min} for downhill slope case, can be easily found.

Water Application Uniformity Criteria

The hydraulic design of a micro-irrigation system to achieve high system uniformity has led design engineers to over-design irrigation systems arbitrarily. In a micro-irrigation system, a major hydraulic design criterion is the minimization of the emitter flow variation, which means the maximization of water application uniformity along a micro-irrigation line (lateral or sub-main). The emitter flow variation is controlled by the pressure head variation along the line, which results from the combined effect of friction drop and uniform line slopes. When the kinetic energy is considered to be small and neglected in a micro-irrigation line, the pressure variation will be simply a linear combination of the energy drop due to friction and energy gain (for downslope) or loss (for upslope) due to uniform line slope [14–18].

There are many expressions to show the uniformity of water application. The uniformity coefficient U_C [48] which evaluates the mean deviation, and a statistical term, the coefficient of variation of emitter discharge (CV_q), which evaluates the standard deviation, are the two most commonly used uniformity expressions [13–15]. Other frequently used uniformity measures are the lower-quarter distribution uniformity (D_U) and emission uniformity (EU), which express a ratio of the minimum to the mean emitter flow [1].

More simple forms for uniformity expression are the emitter flow variation (δ_q) (or pressure head variation, δ_H) along the pipeline [16]. For a micro-irrigation system which has a relatively high uniformity in hydraulic design all the uniformity expressions can be highly correlated to each other, which means the design criterion made for any single uniformity expression (U_C, CV_q, D_U, E_U, δ_q or δ_H) can be converted and used for other uniformity expressions [4]. The constraints for uniformity expressions are essentially based on the hydraulic design of a micro-irrigation system, which can be either to be accepted or rejected. Commonly used emitter flow variations of 10–20% (corresponds to pressure head variations of 20–40%) are equivalent to the uniformity coefficients (U_C or D_U) of about 98–95%, or a coefficient of variation of emitter flow of only 3–7% [14].

Allowable Limit for Extreme Pressure Head Difference

The uniformity criterion applied for the design of multiple outlets pipelines is that the maximum difference in the outlet operating pressure head along the pipeline is less than a percentage of the average pressure head, δ_H:

$$\delta_H = \frac{\Delta H}{\overline{H}} = \frac{H_{\max} - H_{\min}}{\overline{H}} \tag{2.32}$$

where δ_H = percentage of the maximum allowable difference in the outlet operating pressure heads to the average outlet pressure head; ΔH = maximum allowable difference in the outlet operating pressure head along the pipeline; and H_{max} and H_{min} = maximum and minimum values of the operating outlet pressure head, respectively.

Wu [13] proposed the following uniformity parameters given by Eq. (2.33) to alternate Eq. (2.32), which can be specifically used as the design criteria for trickle irrigation submain units. The uniformity criteria δ_H and δ_q (emitter flow variation) along a submain line in trickle irrigation system is given by

$$\delta_H = \frac{\Delta H}{H_{\max}} = 1 - \frac{H_{\min}}{H_{\max}}; \quad \delta_q = \frac{\Delta q}{q_{\max}} = 1 - \frac{q_{\min}}{q_{\max}}; \quad \delta_q = 1 - (1 - \delta_H)^x \tag{2.33}$$

where δq = the emitter flow variation along lateral line or in a submain unit in trickle irrigation system; q_{max} and q_{min} = maximum and minimum values of the emitter outflow, respectively; $\Delta q = q_{max} - q_{min}$ = maximum allowable difference in the emitter outflow along a submain line.

Equation (2.33) indicates δ_q is essentially equivalent to δ_H for the laminar flow emitters ($x = 1.0$). The criteria of hydraulic design are usually set as 10% and 20% emitter outflow variation, δ_q [1], which is equivalent to values of about 20% and 40% of δ_H for an emitter in turbulent flow or sprinkler irrigation system with $x = 0.5$ [13].

The maximum difference in the extreme outlet operating pressure heads can be written in the following general form:

$$\begin{aligned} H_{\max} - H_{\min} = {} & H_{f(L)}\left[(1 - i_{\max})^{m_\beta+1} - (1 - i_{\min})^{m_\beta+1}\right] \\ & + H_{VI}\left[(1 - i_{\min})^2 - (1 - i_{\max})^2\right] - p\Delta S(i_{\max} - i_{\min}) \end{aligned} \tag{2.34}$$

For uphill and zero slope conditions, using the expressions $i_{max} = \mu$ s/L, and i_{min} = 1.0 in Eq. (2.34), then the maximum difference in the outlet operating pressure head is obtained

$$H_{\max} - H_{\min} = H_{f(L)}\theta^{m_\beta+1} - H_{VI}\theta^2 + p\Delta S\theta \tag{2.35}$$

where $\theta = L_0/L$, length ratio which is defined in the previous section.

For downhill slope condition, using the expression for i_{min} from Eq. (2.31), and the value of i_{max} substituting into Eq. (2.34), maximum difference in the outlet operating pressure head can be obtained. Finally, Eq. (2.34) or Eq. (2.35) are divided by $\overline{H}$, then the δ_H can be easily evaluated from Eq. (2.32).

Coefficient of Variation of Discharge (CV_q), Christiansen's Uniformity (U_C) and Lower-Quarter Distribution Uniformity (DU_{LQ})

The variance of pressure head profile between the pipe segments $l = 0$ and $l = L$, can be expressed as [1]

$$H_{\mathrm{var}} = \frac{1}{L}\int_{l=0}^{l=L} DEV_l^2 dl = \frac{1}{L}\int_{l=0}^{l=L}\left[H(l) - \overline{H}\right]^2 dl \tag{2.36}$$

where DEV_l = deviation between $H(l)$ and $\overline{H}$ at any distance l from the pipeline inlet.

Subtracting $\overline{H}$ from Eq. (2.28) and into Eq. (2.36) then, solving the integration yields

$$\begin{aligned}\mathbf{H}_{\mathrm{var}} = {} & \frac{8}{15}H_{VI}^2 + \frac{(m_\beta+1)^2}{(m_\beta+2)^2(2m_\beta+3)}H_{f(L)}^2 - \frac{4}{3}\frac{(m_\beta+1)}{(m_\beta+2)(m_\beta+4)}H_{VI}H_{f(L)} \\ & - \frac{1}{6}H_{VI}(p\Delta S) + \frac{(m_\beta+1)}{(m_\beta+2)(m_\beta+3)}H_{f(L)}(p\Delta S) + \frac{1}{12}(p\Delta S)^2\end{aligned} \tag{2.37}$$

The relationship between CV_q and CV_H is given by the authors [47]

$$CV_q = xCV_H = \frac{x\sqrt{H_{\mathrm{var}}}}{\overline{H}} \tag{2.38}$$

in which CV_q = the coefficient of variation of discharge is statistically defined as the standard deviation of discharge divided by the mean value and expressed in percent; and CV_H = the coefficient of variation of pressure head defined as standard deviation of pressure head divided by the average pressure head and expressed in percent [45]. The commonly used level of the CV_q varies for the intervals 3% and 5% [3, 14].

For normal distribution, the Christiansen's uniformity coefficient U_C is related to CV_q by [53]

$$U_C = 1 - 0.798CV_q \tag{2.39}$$

Substituting Eq. (2.37) for H_{var} into Eq. (2.38), then the CV_q into Eq. (2.39) finally, the U_C coefficient can be directly evaluated from the following Eq. (2.40):

$$U_C = 1 - 0.798\frac{x}{\bar{H}}\begin{pmatrix} \frac{8}{15}H_{VI}^2 + \frac{(m_\beta+1)^2}{(m_\beta+2)^2(2m_\beta+3)}H_{f(L)}^2 - \frac{4}{3}\frac{(m_\beta+1)}{(m_\beta+2)(m_\beta+4)}H_{VI}H_{f(L)} \\ -\frac{1}{6}H_{VI}(p\Delta S) + \frac{(m_\beta+1)}{(m_\beta+2)(m_\beta+3)}H_{f(L)}(p\Delta S) + \frac{1}{12}(p\Delta S)^2 \end{pmatrix}^{1/2} \tag{2.40}$$

For normal distribution, the relationship between the U_C and the lower-quarter distribution uniformity coefficient (DU_{LQ}) is given by the following relationship [38]

$$U_C = 0.37 + 0.63DU_{LQ} \tag{2.41}$$

Substituting the expression for U_C given by Eq. (2.40) into Eq. (2.39) and rearranging for DU_{LQ}, the following simple transformation between the DU_{LQ} and the coefficient of variation of discharge (CV_q) can be deduced:

$$DU_{LQ} = 1 - 1.267CV_q \tag{2.42}$$

Flow Regimes Along the Line

Consider a pipeline with multiple outlets consists of the specific lengths of the segments in which the different flow regimes (i.e., fully turbulent, turbulent, transition and laminar) occur along its length. The variation of flow regime in the specific pipe segments is characterized with decreasing of the Reynolds number in direction of flow [2].

Reynolds Number Variation for Nonuniform Outflow Profile Concept

The variation of the Reynolds number for nonuniform outflow profile exponentially varies along the pipeline in accordance with the flow rate distribution Q(l) as given by Eq. (2.2)

$$R(l) = R_I(1 - i)^\beta \tag{2.43}$$

in which R(l) = Reynolds number for any distance l from the pipeline inlet.

Multiple outlets pipelines are generally made of smooth materials. Flow in a pipeline with multiple outlets is usually turbulent, sometimes fully turbulent flow exists at the upstream end of the pipeline and, flow becomes laminar at the downstream reach where the flow velocity decreases to zero. The transition flow occurs for the interval, $2000 < R \leq 3000$, and the flow characteristics are indeterminate for the transition zone [2].

Determination of Specific Pipe Segments in Different Flow Regimes

In the following, the expressions for the specific pipe segments in different flow regimes are deduced [19] from the definitions of the flow regimes characterized by the specific intervals of the Reynolds number [54].

For fully turbulent flow, the pipe segment in which fully turbulent flow regime prevails along its length can be represented by

$$l_{fully-turbulent} = L \times \left[1 - \left(\frac{10^5}{R_I} \right)^{1/\beta} \right] \tag{2.44}$$

The turbulent flow length for the cases of either fully turbulent flow or no fully turbulent flow exists at the inlet of the pipeline, can be deduced by the following Eqs. (2.45a) and (2.45b), respectively

$$l_{turbulent} = L \times \left[\frac{10^{5/\beta} - 3000^{1/\beta}}{R_I^{1/\beta}} \right] \tag{2.45a}$$

$$l_{turbulent} = L \times \left[1 - \left(\frac{3000}{R_I} \right)^{1/\beta} \right] \tag{2.45b}$$

For the length of the transition zone, the following expression is given by

$$l_{transition} = L \times \left[\frac{3000^{1/\beta} - 2000^{1/\beta}}{R_I^{1/\beta}} \right] \tag{2.46}$$

Although the length of the transition zone can be determined by Eq. (2.46), the current design procedure assumes the most usual case in which the turbulent flow length is practically extended to the start of laminar flow section omitting the transition zone.

Finally, the laminar flow length can be determined as follows

$$l_{la\min ar} = L \times \left[\frac{2000^{1/\beta} - 1}{R_I^{1/\beta}} \right] \tag{2.47}$$

Steps of Algorithm for the Analytical REGL Method

In the analytical procedure presented, the following data are required for a certain design: the internal diameter of the pipe D, the uniform slope along the pipeline S_0, total number of operating outlets N, outlet spacing s, inlet spacing ratio μ, the outlet discharge coefficient and exponent k and x, constants for the Darcy-Weisbach formula (c_1, m and n) given by Eq. (2.5), and finally, the average outlet discharge $\overline{q}$ and corresponding pressure head $\overline{H}$.

It should be noted, the calculation steps of the proposed analytical solution are illustrated in the following computational design examples.

Computational Design Applications

In the following, we illustrate the derivative procedure presented-above, on two computational applications covering trickle and sprinkle irrigation lateral design problems which are originated from the recent work [54].

Design Example 1: Micro-irrigation Lateral Design

Determine the pressure head and outflow profiles along the pipeline for the nonuniform outflow concept and, the corresponding flow characteristics for a horizontal ($S_0 = 0$) polyethylene (PE) trickle irrigation lateral with turbulent flow emitters ($x = 0.54$), using the following data: total number of emitters, $N = 151$ which are equally spaced at $s = 1.0$ m; the inlet spacing ratio from the main pipe, $\mu = s_I/s = 1.0$. The internal diameter of the lateral is considered, $D = 14$ mm. The required average emitter discharge and corresponding nozzle pressure head are, $\overline{q} = 2.0\ \text{lh}^{-1}$ ($= 5.555 \times 10^{-7}\ \text{m}^3\text{s}^{-1}$) and $\overline{H} = 7.2$ m, respectively. Irrigation water is at 20 ^{0}C (kinematic viscosity $\nu = 1.01 \times 10^{-6}\ \text{m}^2\text{s}^{-1}$); and acceleration due to gravity, $g = 9.807\ \text{ms}^{-2}$.

Analytical Solution: Revised Energy-Gradient Line (R-EGL) Method

Before starting the following analytical solution (R-EGL method) for the above design problem, we clearly analyze the input and output parameters, then first identify the type of design problem (Class A, B or C), in accordance with the given and required design parameters.

Input variables:
$S_0 = 0$, $x = 0.54$, $N = 151$, $s = 1.0$ m ($\mu = 1$), $D = 14$ mm, $\overline{q} = 2.0\ \text{lh}^{-1}$ ($= 5.555 \times 10^{-7}\ \text{m}^3\text{s}^{-1}$), $\overline{H} = 7.2$ m, $\nu = 1.01 \times 10^{-6}\ \text{m}^2\text{s}^{-1}$, $g = 9.807\ \text{ms}^{-2}$.

Required parameters:
The pressure head and outflow profiles with the corresponding flow characteristics along the pipeline: Total friction drop at the end of the line, $H_{f(L)}$ and $H_{f(Lo)}$; operating inlet pressure head, H_{0I}; downstream end pressure head, H_N; extreme pressure heads (H_{max},

H_{min}) with their positions (i_{max}, i_{min}); percentage of average pressure head, water application uniformity criteria, δ_H, δ_q, CV_q, CV_H, U_C, $D_{U(LQ)}$, specific length of pipe segments in which different flow regimes are occurred ($l_{fully\text{-}turbulent}$, $l_{turbulent,}$, $l_{transition}$, $l_{laminar}$).

Case (a): for traditional **uniform outflow** concept,

Case (b): for improved **nonuniform outflow** concept using exponential function by Eq. (2.22)

As concluded from the given hydraulic variables and the required parameters, this type of problem can be evaluated in the Class A problems (energy profile determination).

Before starting the analytical solution for **(a)** and **(b)**, the following preliminary steps [Steps 1~8] are first implemented, respectively:

1. The distance between first and last sprinklers $L_0 = (N - 1)s = (151 - 1)\times 1 = 150$ m; and total lateral length $L = (N + \mu - 1)s = (151 + 1 - 1)\times 1 = 151$ m; therefore the length ratio, $\theta = L_0/L = 0.9934$ is computed.
2. Total inflow rate for the entire length of the pipeline and inlet flow velocity are approximated by Eq. (2.2):
   $Q_I = N\overline{q} = 151 \times 5.555 \times 10^{-7} = 8.389 \times 10^{-5}\ m^3s^{-1}$
$$V_I = \frac{4Q_I}{\pi D^2} = \frac{4 \times 8.389 \times 10^{-5}}{\pi \times 0.014^2} = 0.545ms^{-1}$$
3. The total kinetic head at the pipe inlet is evaluated from Eq. (2.4):
$$H_{VI} = \frac{V_I^2}{2g} = \frac{0.545^2}{2 \times 9.81} = 0.015\,\text{m}$$
4. The value of Reynolds number at the pipe inlet is computed from the definition of the Reynolds number, Eq. (2.13):
$$R_I = \frac{4Q_I}{\pi D\nu} = \frac{4 \times 8.389 \times 10^{-5}}{\pi \times 0.014 \times 1.01 \times 10^{-6}} \cong 7{,}554.$$
5. The turbulent flow ($3{,}000 < R_I = 7{,}554 < 10^5$) occurs at the inlet of lateral, therefore the following coefficients/exponents in Eqs. (2.5) and (2.6) are used:
$$c_1 = 0.316,\ \ m = 1.75,\ \ n = 4.75,\ \ C = 7.792 \times 10^{-4}$$
6. The friction coefficient is evaluated from the general expression, Eq. (2.13):
$$f = c_1[R_I]^{m-2} = 0.316 \times (7,554)^{1.75-2} = 0.0339 \cong 0.034$$

7. For the total number of outlets N = 151, the percentage difference between the traditional F and F_C (or F_{SW}) is less than 1% (Fig. 2.2). Therefore, the traditional F (Eq. 2.15) is sufficiently accurate (effect of the number of outlets can be neglected).
   The traditional friction correction factor, F from Eq. (2.15):

$$F = \frac{1}{m+1} = \frac{1}{2.75} = 0.3636$$

8. Total friction drop in a similar pipeline without outlets, H_{f0} from Eq. (2.11):

$$H_{f0} = f\frac{L}{D}H_{VI} = C\frac{Q_I^m}{D^n}L$$

$$H_{f0} = 0.034 \times \frac{151}{0.014} \times 0.015$$

$$= 7.792 \times 10^{-4} \times \frac{(8.389 \times 10^{-5})^{1.75}}{0.014^{4.75}} \times 151 = 5.534\,\text{m}.$$

9. Total friction drop at the end of the pipeline, $H_{f(L)}$ from Eq. (2.10):

$$H_{f(L)} = FH_{f0} = 0.3636 \times 5.534 = 2.012\,\text{m}.$$

Case (a): Pressure head profile for traditional uniform outflow concept [$\beta = 1, m_\beta = m \times \beta = (m = 1.75)$]:

10. Operating inlet pressure head, H_{0I}, is directly determined from Eq. (2.29) [or from Eq. (2.21) for the pressure head profile for i = 0]:

$$H_{0I} = \overline{H} + \frac{(m+1)}{(m+2)}H_{f(L)} + p\frac{\Delta S}{2} = 7.2 + \frac{2.75}{3.75} \times 2.012 + 0 = 8.675 \cong 8.68m$$

11. The maximum pressure head, H_{max} from Eq. (2.30):

$$H_{max} = \overline{H} + H_{VI}(1-\theta^2) + H_{f(L)}\left[\theta^{m+1} - \frac{1}{(m+2)}\right] + p\Delta S\left(\theta - \frac{1}{2}\right)$$

$$H_{max} = 7.2 + 0.015 \times (1 - 0.9934^2)$$

$$+ 2.012 \times \left[0.9934^{2.75} - \frac{1}{3.75}\right] + 0 = 8.639 \cong 8.64\,\text{m}$$

12. For horizontal pipe slope situation (p = 0), the minimum pressure head, H_{min}, is at the closed end of the line, and is essentially equal to the downstream pressure head ($H_N = H_{min}$), and evaluated from Eq. (2.21) as follows

$$H_N = H_{min} = \overline{H} + H_{VI} - \left(\frac{1}{m+2}\right)H_{f(L)} - \frac{p\Delta S}{2}$$

$$H_N = H_{min} = 7.2 + 0.015 - \frac{1}{3.75} \times 2.012 - 0 = 6.678\,\text{m}$$

13. The total friction drop for the pipe length (L_0) $H_{f(LO)}$, from the conservation of energy principle, between the first and last emitters:

$$H_{f(L_O)} \cong H_{max} + H_{VI} - H_N = 8.639 + 0.015 - 6.678 = 1.976\,\text{m}$$

$< H_{f(L)} = 2.012$ m.

14. The percentage of average pressure head, δ_H is evaluated by substituting Eq. (2.35) into Eq. (2.32):

$$\delta_H = \frac{\Delta H}{\overline{H}} = \frac{(H_{max} - H_{min})}{\overline{H}} \cong \theta^{m+1}\frac{H_{f(L)}}{\overline{H}} - \theta^2\frac{H_{VI}}{\overline{H}} + \theta\frac{(p\Delta S)}{\overline{H}}$$

$$\delta_H = \frac{H_{max} - H_{min}}{\overline{H}} = \frac{(8.639 - 6.678)}{7.2} = 0.27 = 27\% \text{ or}$$

$$\delta_H \cong 0.9934^{2.75} \times \frac{2.012}{7.2} - 0.9934^2 \times \frac{0.015}{7.2} + 0.9934 \times 0 = 0.27 = 27\%$$

15. Then, the emitter flow variation, δ_q is evaluated from the relationship between δ_H and δ_q, as given by Eq. (2.33):

$$\delta_q = 1 - (1 - \delta_H)^x = 1 - (1 - 0.26)^{0.54} \cong 0.16 = 16\%$$

16. The variance of pressure head profile between the pipe segments $l = 0$ and $l = \text{L}$, can be evaluated from the analytical expression, Eq. (2.37), for $m_\beta = m = 1.75$:

$$\text{H}_{\text{var}} = \frac{8}{15}H_{VI}^2 + \frac{(m+1)^2}{(m+2)^2(2m+3)}H_{f(L)}^2 - \frac{4}{3}\frac{(m+1)}{(m+2)(m+4)}H_{VI}H_{f(L)}$$
$$-\frac{1}{6}H_{VI}(p\Delta S) + \frac{(m+1)}{(m+2)(m+3)}H_{f(L)}(p\Delta S) + \frac{1}{12}(p\Delta S)^2$$

$$\text{H}_{\text{var}} = \frac{8}{15} \times 0.015^2 + \frac{2.75^2}{3.75^2(2 \times 1.75 + 3)} \times 2.012^2 - \frac{4}{3} \times \frac{2.75}{3.75 \times 5.75} \times 0.015$$
$$\times 2.012 - \frac{1}{6} \times 0.015 \times (0)$$
$$+ \frac{2.75}{3.75 \times 4.75} \times 2.012 \times (0) + \frac{1}{12} \times (0)^2 \cong 0.33 = 33\%$$

17. The coefficient of variation of emitter outflow, CV_q is directly evaluated from Eq. (2.38):

$$CV_q = \frac{x\sqrt{H_{var}}}{\overline{H}} = \frac{0.54 \times \sqrt{0.33}}{7.2} = 0.043 = 4.3\%$$

Then, the coefficient of variation of pressure head, CV_H, from the relationship between CV_q and CV_H, as given by Eq. (2.38):

$$CV_H = \frac{CV_q}{x} = \frac{0.043}{0.54} = 0.0798 \cong 8\%$$

18. The Christiansen's uniformity coefficient, U_C is evaluated from the analytical expression, Eq. (2.40), or directly from the relationship between the U_C and CV_q [Eq. (2.39)]:

$$U_C = 1 - 0.798CV_q = 1 - 0.798 \times 0.043 = 0.966 = 96.6\%$$

19. The lower-quarter distribution uniformity, DU_{LQ} is evaluted from the relationship between U_C and DU_{LQ} [Eq. (2.41)] or the relationship between DU_{LQ} and CV_q [Eq. (2.42)]:

$$U_C = 0.37 + 0.63DU_{LQ}$$

$$DU_{LQ} = \frac{(U_C - 0.37)}{0.63} = \frac{(0.966 - 0.37)}{0.63} = 0.945 = 94.5\%$$

$$DU_{LQ} = 1 - 1.267CV_q = 1 - 1.267 \times 0.043 = 0.945 = 94.5\%$$

20. The turbulent flow ($3{,}000 < R_I = 7{,}554 < 10^5$) occurs at the inlet of lateral, as illustrated in Step-4, so the lengths of specific pipe segments in which the turbulent flow, transition flow and laminar flow occur along its length, $l_{turbulent}$, $l_{transition}$ and $l_{laminar}$, can be determined with considering the nonuniform outflow exponent, $\beta = 1$, for uniform outflow profile:

(a) For the flow case of no fully-turbulent flow is occurred at the lateral inlet, Eq. (2.45b) is directly used for the turbulent flow length, $l_{\text{turbulent}}$:

$$l_{turbulent} = L \times \left[1 - \left(\frac{3,000}{R_I}\right)^{1/\beta}\right]$$

$$= 151 \times \left[1 - \left(\frac{3,000}{7,554}\right)^{1}\right] = 91.0m$$

$$i_{\text{turbulent}} = \frac{l_{turbulent}}{L} = \frac{91}{151} = 0.603 \cong 60\% \triangleleft$$

(b) For the length of the transition zone, Eq. (2.46) is used:

$$l_{transition} = L \times \left[\frac{3,000^{1/\beta} - 2,000^{1/\beta}}{R_I^{1/\beta}}\right]$$

$$= 151 \times \left[\frac{3,000^1 - 2,000^1}{7,554^1}\right] \cong 20.0m$$

$$i_{transition} = \frac{l_{transition}}{L} = \frac{20}{151} = 0.132 \cong 13.2\%$$

(c) Finally, the laminar flow length can be determined from Eq. (2.47):

$$l_{la\min ar} = L \times \left[\frac{2,000^{1/\beta} - 1}{R_I^{1/\beta}}\right]$$

$$= 151 \times \left[\frac{2,000^1 - 1}{7,554^1}\right] = 39.96 \cong 40m$$

$$i_{laminar} = \frac{l_{laminar}}{L} = \frac{40}{151} = 0.265 \cong 26.5\%$$

In the rest of calculation, this analytical procedure (R-EGL method) will be continued by the following improved stage by repeating the previous steps in **(a)**: ***Steps 5~18***, for **(b)**: the improved nonuniform outflow profile based on the exponential function [Eq. (2.22)].

Case (b): Pressure head profile for improved nonuniform outflow concept

[$\beta \neq 1, m_\beta = m \times \beta \neq (m = 1.75)$]:

1. The nonuniform outflow exponent, β is determined by Eq. (2.27):

First, the parameter, λ in the expression of β, is computed for m = 1.75:

$$\lambda = \frac{1 - 0.5^{m+1}}{m+2} = \frac{1 - 0.5^{2.75}}{3.75} = 0.027$$

For the previously known/computed values of parameters;
$H_{VI} = 0.015$ m, $H_{f(L)} = 2.012$ m, $\overline{H} = 7.2$ m, $x = 0.54$,
p = 0 (Horizontal pipe case, $\Delta S = S_0 \times L = 0$)

$$\beta = 1 - 3.322x \log\left[1 + \frac{H_{VI}}{\overline{H}} + \lambda \frac{H_{f(L)}}{\overline{H}} + \frac{1}{4\overline{H}}(p\Delta S)\right]$$

$$\beta = 1 - 3.322 \times 0.54 \times \log\left[1 + \frac{0.015}{7.2} + 0.227 \times \frac{2.012}{7.2} + \frac{1}{4 \times 7.2} \times (0)\right] = 0.9506 \cong 0.951$$

2. The improved value of velocity exponent, m_β:

$$m_\beta = m \times \beta = 1.75 \times 0.951 = 1.664$$

3. The traditional friction correction factor, F from Eq. (2.15) [*(a): Step 6*]:

$$F = \frac{1}{m_\beta + 1} = \frac{1}{(1.664 + 1)} = 0.3753$$

4. Total friction drop at the end of the pipeline, $H_{f(L)}$ [*(a): Step 8*]:

$H_{f(L)} = FH_{f0} = 0.3753 \times 5.534 = 2.08$ m.

5. Operating inlet pressure head, H_{0I}, from Eq. (2.29) [*(a): Step 9*]:

$$H_{0I} = \overline{H} + \frac{(m_\beta + 1)}{(m_\beta + 2)} H_{f(L)} + p\frac{\Delta S}{2} = 7.2 + \frac{(1.664 + 1)}{(1.664 + 2)} \times 2.08 + 0 = 8.712 \cong 8.71m$$

6. The maximum pressure head, H_{max} from Eq. (2.30): [*(a): Step 10*]:

$$H_{max} = \overline{H} + H_{VI}(1-\theta^2) + H_{f(L)}\left[\theta^{m_\beta+1} - \frac{1}{(m_\beta+2)}\right] + p\Delta S\left(\theta - \frac{1}{2}\right)$$

$$\begin{aligned} H_{\max} &= 7.2 + 0.015 \times (1 - 0.9934^2) \\ &+ 2.08 \times \left[0.9934^{(1.664+1)} - \frac{1}{(1.664+2)}\right] + 0 = 8.676 \cong 8.68\,\text{m} \end{aligned}$$

7. The minimum pressure head, H_{min}, which is equal to the downstream pressure head ($H_N = H_{min}$), from Eq. (2.21) [*(a): Step 11*]:

$$H_N = H_{min} = \overline{H} + H_{VI} - \left(\frac{1}{m_\beta+2}\right)H_{f(L)} - \frac{p\Delta S}{2}$$

$$H_N = H_{min} = 7.2 + 0.015 - \frac{1}{(1.664+2)} \times 2.08 - 0 = 6.647 \cong 6.65m$$

8. The total friction drop for the pipe length (L_0) $H_{f(LO)}$, from the conservation of energy principle [*(a): Step 12*]:

$$H_{f(L_O)} \cong H_{max} + H_{VI} - H_N = 8.676 + 0.015 - 6.647 = 2.04m < H_{f(L)}2.08m.$$

9. The percentage of average pressure head, δ_H is evaluated by substituting Eq. (2.35) into Eq. (2.32) [*(a): Step 13*]:

$$\delta_H = \frac{\Delta H}{\overline{H}} = \frac{(H_{max} - H_{min})}{\overline{H}} \cong \theta^{m_\beta+1}\frac{H_{f(L)}}{\overline{H}} - \theta^2\frac{H_{VI}}{\overline{H}} + \theta\frac{(p\Delta S)}{\overline{H}}$$

$$\delta_H = \frac{H_{max} - H_{min}}{\overline{H}} = \frac{(8.676 - 6.647)}{7.2} = 0.28 = 28\%$$

$$\delta_H \cong 0.9934^{(1.664+1)} \times \frac{2.08}{7.2} - 0.9934^2 \times \frac{0.015}{7.2} + 0.9934 \times 0 = 0.28 = 28\%$$

10. The emitter flow variation, δ_q from Eq. (2.33) [*(a): Step 14*]:

$$\delta_q = 1 - (1 - \delta_H)^x = 1 - (1 - 0.28)^{0.54} = 0.16 = 16\%$$

11. The variance of pressure head profile H_{var}, can be evaluated from Eq. (2.37), for $m_\beta = 1.664$ [*(a): Step 15*]:

$$\mathrm{H_{var}} = \frac{8}{15}H_{VI}^2 + \frac{(m_\beta+1)^2}{(m_\beta+2)^2(2m_\beta+3)}H_{f(L)}^2 - \frac{4}{3}\frac{(m_\beta+1)}{(m_\beta+2)(m_\beta+4)}H_{VI}H_{f(L)} - \frac{1}{6}H_{VI}(p\Delta S) + \frac{(m_\beta+1)}{(m_\beta+2)(m_\beta+3)}H_{f(L)}(p\Delta S) + \frac{1}{12}(p\Delta S)^2$$

$$\mathrm{H_{var}} = \frac{8}{15}\times 0.015^2 + \frac{(1.664+1)^2}{(1.664+2)^2\times(2\times 1.664+3)} \times 2.08^2 - \frac{4}{3}\times\frac{(1.664+1)}{(1.664+2)\times(1.664+4)}\times 0.015\times 2.08 - \frac{1}{6}\times 0.015\times(0) + \frac{(1.664+1)}{(1.664+2)\times(1.664+3)}\times 2.08\times(0) + \frac{1}{12}\times(0)^2 = 0.356 \cong 36\%$$

12. The coefficient of variation of emitter outflow, CV_q from Eq. (2.38) [*(a): Step 16*]:

$$CV_q = \frac{x\sqrt{H_{\mathrm{var}}}}{\overline{H}} = \frac{0.54\times\sqrt{0.356}}{7.2} = 0.045 = 4.5\%$$

Then, the coefficient of variation of pressure head, CV_H, from Eq. (2.38):

$$CV_H = \frac{CV_q}{x} = \frac{0.045}{0.54} = 0.083 \cong 8.3\%$$

13. The Christiansen's uniformity coefficient, U_C from Eq. (2.39) [*(a): Step 17*]:

$$U_C = 1 - 0.798CV_q = 1 - 0.798\times 0.045 = 0.964 = 96.4\%$$

14. The lower-quarter distribution uniformity, DU_{LQ} is evaluted from the relationship between U_C and DU_{LQ} [Eq. (2.41)] or the relationship between DU_{LQ} and CV_q [Eq. (2.42)] [*(a): Step 18*]:

$$U_C = 0.37 + 0.63DU_{LQ}$$

$$DU_{LQ} = \frac{(U_C - 0.37)}{0.63} = \frac{(0.964-0.37)}{0.63} = 0.943 = 94.3\%$$

$$DU_{LQ} = 1 - 1.267CV_q = 1 - 1.267 \times 0.045 = 0.943 = 94.3\%$$

15. The lengths of specific pipe segments in which the turbulent flow, transition flow and laminar flow prevail along its length, $l_{\text{turbulent}}$, $l_{\text{transition}}$ and l_{laminar}, can be determined with considering the nonuniform outflow exponent, $\beta = 0.951$, for nonuniform outflow profile [*(a): Step 19*]:

(a) *For the flow case of no fully-turbulent flow is occurred at the lateral inlet*

($3{,}000 < \mathrm{R_I} = 7{,}554 < 10^5$), Eq. (2.45b) is directly used for the turbulent flow length, $l_{\text{turbulent}}$:

$$l_{turbulent} = L \times \left[1 - \left(\frac{3,000}{R_I}\right)^{1/\beta}\right]$$

$$= 151 \times \left[1 - \left(\frac{3,000}{7,554}\right)^{1/0.951}\right] = 93.8 \cong 94.0m$$

$$i_{\text{turbulent}} = \frac{l_{turbulent}}{L} = \frac{94}{151} = 0.621 \cong 62\%$$

(b) ***For the length of the transition zone***, Eq. (2.46) is used:

$$l_{transition} = L \times \left[\frac{3,000^{1/\beta} - 2,000^{1/\beta}}{R_I^{1/\beta}}\right]$$

$$= 151 \times \left[\frac{3,000^{1/0.951} - 2,000^{/0.9511}}{7,554^{/0.9511}}\right] = 19.85 \cong 20.0m$$

$$i_{transition} = \frac{l_{transition}}{L} = \frac{20}{151} = 0.134 \cong 13.4\%$$

(c) Finally, ***the laminar flow*** length can be determined from Eq. (2.47):

$$l_{laminar} = L \times \left[\frac{2,000^{1/\beta} - 1}{R_I^{1/\beta}}\right]$$

$$= 151 \times \left[\frac{2,000^{1/0.951} - 1}{7,554^{1/0.951}}\right] = 37.32 \cong 37.3m$$

$$i_{\text{laminar}} = \frac{l_{laminar}}{L} = \frac{37}{151} = 0.247 \cong 24.7\%$$

For the sake of comparison of the complete results obtained from the REGL method for the design cases **(a): uniform outflow profile**, and **(b): nonuniform outflow profile**, are clearly synthesized in Table 2.1. The same example was also extended to cover various uniform slope conditions (S_0 = 0.0, −0.02, −0.05, 0.02 and, 0.05) and, results were shown in Fig. 2.3. In this figure, the pressure head and outflow profiles in comparison with the average values ($\overline{H} = 7.2m$, $\overline{q} = 5.555 \times 10^{-7} m^3 s^{-1}$) versus the dimensionless distance (i = l/L) were also presented in Fig. 2.3a,b, respectively. As a remarkable result from Fig. 2.3a,b, for different slope values the pressure head (or outflow) profile roughly intersects in the middle of the length of the lateral and, its value in this point (i = 0.5) is approximately equal to the average values.

Example 1 (trickle lateral design) for **(a):** uniform outflow profile (β = 1) and **(b):** nonuniform outflow profile (β = 0.951).

Design Example 2: Sprinkler Lateral Design

Determine the energy grade-line (EGL) along the pipeline with the corresponding flow characteristics for a horizontal (S_0 = 0) sprinkler irrigation lateral (x = 0.5), using given the following data: total number of sprinklers, N = 33 which are equally spaced at s = 12. 0 m; the inlet spacing ratio from the main pipe, μ = s_l/s = 1.0. Lateral pipe is of aluminum (equivalent roughness, ε = 0.1 mm); irrigation water is at 20 °C (kinematic viscosity, ν = 1.01 × 10^{-6} m^2s^{-1}); and acceleration due to gravity, g = 9.807 ms^{-2}. The required average sprinkler discharge and corresponding nozzle pressure head are, $\overline{q}$ = 0.315 ls^{-1} and $\overline{H}$ = 31.61 m, respectively. The internal diameter of the lateral is considered, D = 101.6 mm (≅ 4 in). Consider neglecting minor head losses in sprinkler risers except for the energy losses due to couplers. The friction loss multiplier is taken 1.08 for energy losses in couplers for the value of the common outlet spacing s = 12 m [10].

Analytical Solution: Revised Energy-Gradient Line (R-EGL) Method

Before starting the following analytical solution (R-EGL method) for the above design problem, we clearly analyze the input and output parameters, then first identify the type of design problem (Class A, B or C), in accordance with the given and required design parameters.

Input variables:

S_0 = 0, x = 0.5, N = 33, s = 12.0 m (μ = 1), aluminum pipe (equivalent roughness, ε = 0.1 mm); D = 101.6 mm (≅ 4 in), $\overline{q}$ = 0.315 ls^{-1} (= 3.15 × 10^{-4} m^3s^{-1}), $\overline{H}$ = 31.61 m, multiplier for coupler head loss: 1.08, ν = 1.01 × 10^{-6} m^2s^{-1}, g = 9.807 ms^{-2}.

Table 2.1 Comparison of results for hydraulic variables computed from the REGL method using input parameters given in Design Example 1 (trickle lateral design) for (**a**): uniform outflow profile ($\beta = 1$) and (**b**): nonuniform outflow profile ($\beta = 0.951$)

	Results Obtained from the REGL Method for:																	
	Pressure Parameters:						Uniformity Parameters:							Flow Regime Parameters:					
	F	$H_{f(L)}$	$H_{f(Lo)}$	H_{0l}	H_{max}	$H_{min} = H_N$	δ_H	δ_q	H_{var}	CV_q	CV_H	U_C	DU_{LQ}	$l_{turbulent}$	$l_{transition}$	$l_{laminar}$	$i_{turbulent}$	$i_{transition}$	$i_{laminar}$
	(—)	(m)	(m)	(m)	(m)	(m)	(%)	(%)	(%)	(%)	(%)	(%)	(%)	(m)	(m)	(m)	(%)	(%)	(%)
Equation number:	*(4.15)*	*(4.10)*	*(4.10)*	*(4.21)* *(4.29)*	*(4.30)*	*(4.21)*	*(4.32)* *(4.35)*	*(4.33)*	*(4.37)*	*(4.38)*	*(4.38)*	*(4.39)* *(4.40)*	*(4.41)* *(4.42)*	*(4.45b)*	*(4.46)*	*(4.47)*	(—)	(—)	(—)
(**a**): **Uniform** Outflow Profile	0.3636	2.012	1.976	8.675	8.639	6.678	26	16	33	4.3	8.0	96.6	94.5	91	20	40	60	13	27
(**b**): **Nonuniform** Outflow Profile	0.3753	2.08	2.04	8.712	8.676	6.647	28	16	36	4.5	8.3	96.4	94.3	94	20	37	62	13	25

(**a**): **Uniform** Outflow Profile ($\beta = 1$, m = 1.75) and (**b**): **Nonuniform** Outflow Profile ($\beta = 0.951$, $m_\beta = 1.664$)

$\overline{H} = 7.2$ m, $\theta = L_0/L = 0.9934$, $x = 0.54$, p = 0 (Horizontal pipe case), $Q_l = 8.389 \times 10^{-5}$ m^3s^{-1}, $H_{Vl} = 0.015$ m, $R_l = 7{,}554$, $H_{f0} = 5.534$ m,

Turbulent flow situation ($3{,}000 < R_l = 7{,}554 < 10^5$), $f = 0.034$ ($c_1 = 0.316$, m = 1.75, n = 4.75, $C = 7.792 \times 10^{-4}$)

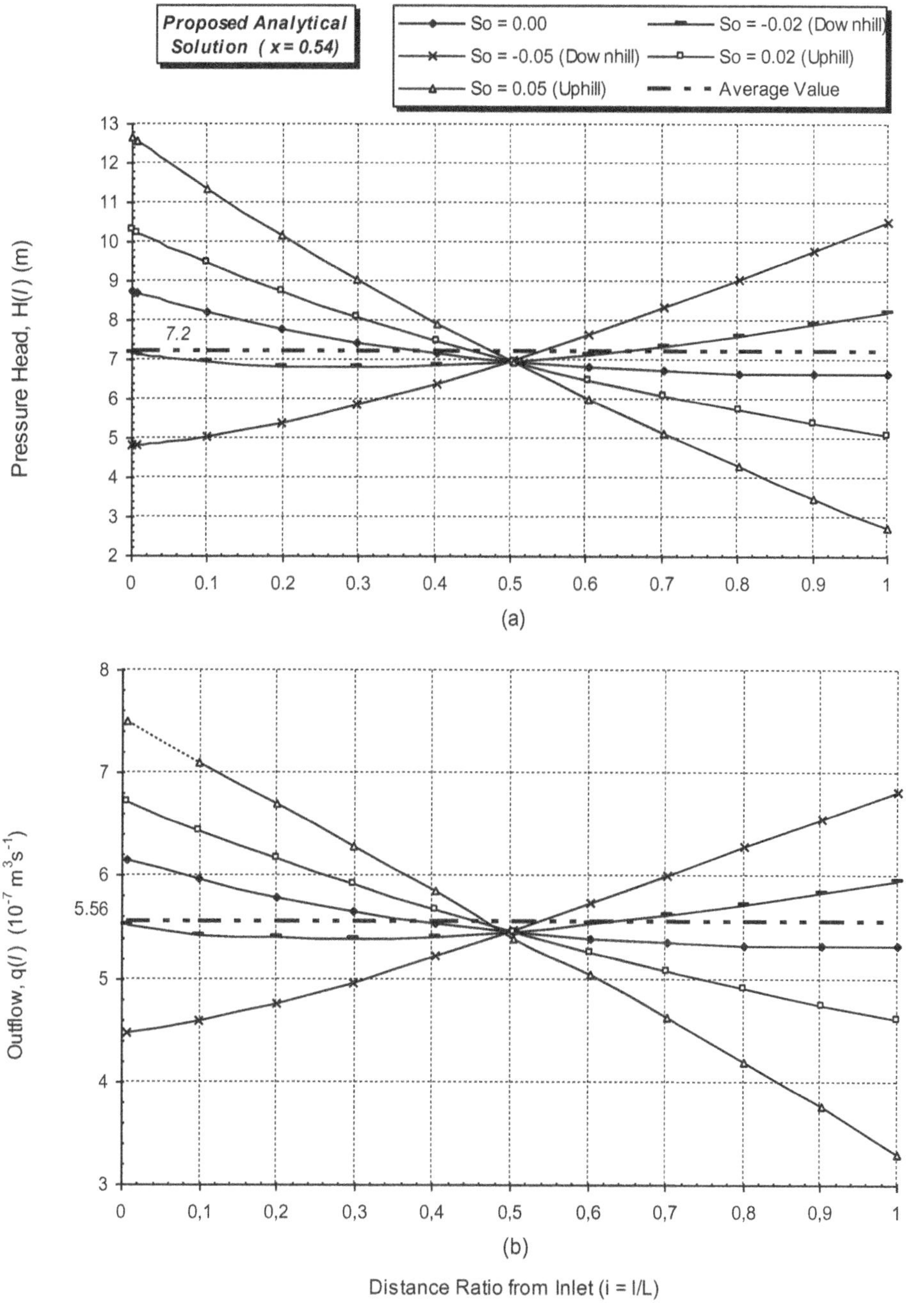

Fig. 2.3 **a** Pressure head and **b** outflow profiles with respect to the distance ratio from inlet, for various uniform line slope cases ($S_O = 0.0, -0.02, -0.05, 0.02$ and 0.05) and for the emitter discharge exponent ($x = 0.54$), according to the presented analytical solution (REGL Method) using given data in Design Example 1

Required parameters:

The pressure head and outflow profiles with the corresponding flow characteristics along the pipeline: Total friction drop at the end of the line, $H_{f(L)}$ and $H_{f(Lo)}$; operating inlet pressure head, H_{0I}; downstream end pressure head, H_N; extreme pressure heads (H_{max}, H_{min}) with their positions (i_{max}, i_{min}); percentage of average pressure head, water application uniformity criteria, δ_H, δ_q, CV_q, CV_H, U_C, $D_{U(LQ)}$, specific length of pipe segments in which different flow regimes are occurred ($l_{fully\text{-}turbulent}$, $l_{turbulent,}$, $l_{transition}$, $l_{laminar}$).

(a) *for traditional uniform outflow concept,*
(b) *for improved nonuniform outflow concept using exponential function by Eq.* ***(2.22)***

As concluded from the given hydraulic variables and the required parameters, this type of problem can be also evaluated in the Class A problems (energy profile determination).

Before starting the analytical solution for **(a)** and **(b)**, the following preliminary steps [Steps 1~8] are first implemented, respectively:

1. The distance between first and last sprinklers $L_0 = (N - 1)s = (33 - 1)\times 12 = 384$ m; and total lateral length $L = (N + \mu - 1)s = (33 + 1\text{–}1)\times 12 = 396$ m; therefore the length ratio, $\theta = L_0/L = 0.9697$ is computed.
2. Total inflow rate for the entire length of the pipeline is approximated from Eq. (2.2):

$$Q_I = N\bar{q} = 33 \times 3.15 \times 10^{-4} = 1.04 \times 10^{-2}\,\text{m}^3\,\text{s}^{-1}$$

3. The total kinetic head at the pipe inlet is evaluated from Eq. (2.4):

$$H_{VI} = \frac{8Q_I^2}{\pi^2 g D^4} = \frac{8 \times (1.04 \times 10^{-2})^2}{\pi^2 \times 9.81 \times (101.6 \times 10^{-3})^4} \cong 0.084\text{m}.$$

4. The value of Reynolds number at the pipe inlet is computed from the definition of the Reynolds number, Eq. (2.13):

$$R_I = \frac{4Q_I}{\pi D \nu} = \frac{4 \times 1.04 \times 10^{-2}}{\pi \times (101.6 \times 10^{-3}) \times 1.01 \times 10^{-6}} \cong 129,041$$

5. The fully-turbulent flow ($10^5 < R_I = 129{,}041 < 10^7$) occurs at the sprinkler lateral inlet, and for aluminum pipes, the friction coefficient f is given by the well-known Churchill equation [5], as given by Eq. (2.7):

$$f = 8\left[\left(\frac{8}{R_I}\right)^{12} + \frac{1}{(A+B)^{3/2}}\right]^{1/12}$$

First, the empirical parameters A and B should be determined from Eq. (2.8) as follows:

$$A = \left\{2.457\ln\left[\frac{1}{\left(\left(\frac{7}{R_I}\right)^{0.9} + 0.27\left(\frac{\varepsilon}{D}\right)\right)}\right]\right\}^{16}$$

$$= \left\{2.457 \times \ln\left[\frac{1}{\left(\left(\frac{7}{129{,}041}\right)^{0.9} + 0.27 \times \left(\frac{0.1\times10^{-3}}{101.6\times10^{-3}}\right)\right)}\right]\right\}^{16} = 3.297 \times 10^{20}$$

$$B = \left(\frac{37,530}{R_I}\right)^{16} = 2.621 \times 10^{-9}$$

Finally, the friction coefficient, f :

$$f = 8\left[\left(\frac{8}{R_I}\right)^{12} + \frac{1}{(A+B)^{3/2}}\right]^{1/12}$$

$$= 8 \times \left[\left(\frac{8}{129,041}\right)^{12} + \frac{1}{(3.297 \times 10^{20} + 2.621 \times 10^{-9})^{3/2}}\right]^{1/12} \cong 0.0218$$

6. For the total number of outlets N = 33, the percentage difference between the F and F_C (or F_{SW}) is about 4.0% (Fig. 2.2). As a preference Eq. (2.16) is taken into account into Eq. (2.12), instead of the traditional Eqs. (2.10)–(2.15).

The Christiansen's friction correction factor, F_C for $\mu = s_I/s = 1.0$, from Eq. (2.16):

$$F_C = \frac{1}{(m+1)} + \frac{1}{2N} + \frac{\sqrt{m-1}}{6N^2}$$

$$F_C = \frac{1}{(1.828+1)} + \frac{1}{2 \times 33} + \frac{\sqrt{1.828-1}}{6 \times 33^2} = 0.3689$$

7. The head loss for the couplers is considered by increasing friction losses by 8%; resulting total friction drop in a similar pipeline without outlets, H_{f0} by using Eq. (2.11):

$$H_{f0} = 1.08 f \frac{L}{D} H_{VI}$$

$$H_{f0} = 1.08 \times 0.0218 \times \frac{396}{0.1016} \times 0.084 = 7.708 \cong 7.71\text{m}.$$

8. Total friction drop at the end of the pipeline, $H_{f(L)}$ from Eq. (2.10):

$$H_{f(L)} = F_C H_{f0} = 0.3689 \times 7.708 = 2.843\text{m}.$$

Case (a): Pressure head profile for traditional uniform outflow concept [$\beta = 1, m_\beta = m \times \beta = (m = 1.828)$]:

9. Operating inlet pressure head, H_{0I}, from Eq. (2.29) [or from Eq. (2.21) for the pressure head profile for: i = 0]:

$$H_{0I} = \overline{H} + \frac{(m+1)}{(m+2)} H_{f(L)} + p\frac{\Delta S}{2} = 31.61 + \frac{2.828}{3.828} \times 2.843 + 0 = 33.71m$$

10. The maximum pressure head, H_{max} from Eq. (2.30):

$$H_{max} = \overline{H} + H_{VI}(1-\theta^2) + H_{f(L)}\left[\theta^{m+1} - \frac{1}{(m+2)}\right] + p\Delta S\left(\theta - \frac{1}{2}\right)$$

$$\begin{aligned} H_{max} &= 31.61 + 0.084 \times (1 - 0.9697^2) + 2.843 \\ &\times \left[0.9697^{2.828} - \frac{1}{3.828}\right] + 0 = 33.481\text{ m} \end{aligned}$$

11. The minimum pressure head, H_{min}, which is essentially equal to the downstream pressure head ($H_N = H_{min}$), evaluated from Eq. (2.21):

$$H_N = H_{min} = \overline{H} + H_{VI} - \left(\frac{1}{m+2}\right) H_{f(L)} - \frac{p\Delta S}{2}$$

$$H_N = H_{min} = 31.61 + 0.084 - \frac{1}{3.828} \times 2.843 - 0 = 30.956m$$

12. The total friction drop for the pipe length (L_0) $H_{f(L0)}$, from the conservation of energy principle, between the first and last emitters:

$$H_{f(L_O)} \cong H_{max} + H_{VI} - H_N = 33.481 + 0.084 - 30.956 = 2.609m < H_{f(L)} = 2.843 \text{ m}.$$

13. The percentage of average pressure head, δ_H is evaluated by substituting Eq. (2.35) into Eq. (2.32):

$$\delta_H = \frac{\Delta H}{\overline{H}} = \frac{(H_{max} - H_{min})}{\overline{H}} \cong \theta^{m+1}\frac{H_{f(L)}}{\overline{H}} - \theta^2\frac{H_{VI}}{\overline{H}} + \theta\frac{(p\Delta S)}{\overline{H}}$$

$$\delta_H = \frac{H_{max} - H_{min}}{\overline{H}} = \frac{(33.481 - 30.956)}{31.61} = 0.0798 \cong 8\% \text{ or}$$

$$\delta_H \cong 0.9697^{2.828} \times \frac{2.843}{31.61} - 0.9697^2 \times \frac{0.084}{31.61} + 0.9697 \times 0 = 0.0798 \cong 8\%$$

14. Then, the sprinkler flow variation, δ_q is evaluated from the relationship between δ_H and δ_q, as given by Eq. (2.33):

$$\delta_q = 1 - (1 - \delta_H)^x = 1 - (1 - 0.0798)^{0.5} \cong 0.041 = 4.1\%$$

15. The variance of pressure head profile, H_{var}, from Eq. (2.37), for $m_\beta = m = 1.852$:

$$\begin{aligned}H_{var} = &\frac{8}{15}H_{VI}^2 + \frac{(m+1)^2}{(m+2)^2(2m+3)}H_{f(L)}^2\\ &- \frac{4}{3}\frac{(m+1)}{(m+2)(m+4)}H_{VI}H_{f(L)} - \frac{1}{6}H_{VI}(p\Delta S)\\ &+ \frac{(m+1)}{(m+2)(m+3)}H_{f(L)}(p\Delta S) + \frac{1}{12}(p\Delta S)^2\end{aligned}$$

$$\begin{aligned}H_{var} = &\frac{8}{15} \times 0.084^2 + \frac{2.828^2}{3.828^2(2 \times 1.828 + 3)} \times 2.843^2\\ &- \frac{4}{3} \times \frac{2.828}{3.828 \times 5.828} \times 0.084 \times 2.843 - \frac{1}{6} \times 0.084 \times (0)\\ &+ \frac{2.828}{3.828 \times 4.828} \times 2.843 \times (0) + \frac{1}{12} \times (0)^2 \cong 0.624 = 62.4\%\end{aligned}$$

16. The coefficient of variation of sprinkler outflow, CV_q is directly evaluated from Eq. (2.38):

$$CV_q = \frac{x\sqrt{H_{var}}}{\overline{H}} = \frac{0.5 \times \sqrt{0.624}}{31.61} = 0.0125 = 1.25\%$$

Then, the coefficient of variation of pressure head, CV_H, from the relationship between CV_q and CV_H, as given by Eq. (2.38):

$$CV_H = \frac{CV_q}{x} = \frac{0.0125}{0.5} = 0.025 \cong 2.5\%$$

17. The Christiansen's uniformity coefficient, U_C is evaluated from the analytical expression, Eq. (2.40), or directly from the relationship between the U_C and CV_q [Eq. (2.39)]:

$$U_C = 1 - 0.798CV_q = 1 - 0.798 \times 0.0125 = 0.99 = 99\%$$

18. The lower-quarter distribution uniformity, DU_{LQ} is evaluted from the relationship between U_C and DU_{LQ} [Eq. (2.41)] or the relationship between DU_{LQ} and CV_q [Eq. (2.42)]:

$$U_C = 0.37 + 0.63DU_{LQ}$$

$$DU_{LQ} = \frac{(U_C - 0.37)}{0.63} = \frac{(0.99 - 0.37)}{0.63} = 0.984 = 98.4\%$$

$$DU_{LQ} = 1 - 1.267CV_q = 1 - 1.267 \times 0.0125 = 0.984 = 98.4\%$$

19. The fully-turbulent flow ($10^5 < R_I = 129{,}041 < 10^7$) occurs at the inlet of sprinkler lateral, as illustrated in ***Step 5***, so the lengths of specific pipe segments in which the fully-turbulent flow, turbulent flow, transition flow and laminar flow prevail along

its length, $l_{\text{turbulent}}$, $l_{\text{transition}}$ and l_{laminar}, can be determined with considering the nonuniform outflow exponent, $\beta = 1$, for uniform outflow profile:

20.

(a) The length of pipe inlet section in which fully turbulent flow occurs, is directly determined from Eq. (2.44):

$$l_{fully-turbulent} = L \times \left[1 - \left(\frac{10^5}{R_I}\right)^{1/\beta}\right]$$

$$= 396 \times \left[1 - \left(\frac{10^5}{129,041}\right)^{1/1}\right] = 89.1 \cong 89m$$

$$i_{fully-\text{turbulent}} = \frac{l_{fully-turbulent}}{L} = \frac{89}{396} = 0.225 \cong 22.5\%$$

(b) The turbulent flow length, $l_{\text{turbulent}}$ for the case of fully-turbulent flow exists at the inlet of the pipeline, can be determined Eq. (2.45a):

$$l_{turbulent} = L \times \left[\frac{10^{5/\beta} - 3000^{1/\beta}}{R_I^{1/\beta}}\right]$$

$$= 396 \times \left[\frac{10^{5/1} - 3,000^{1/1}}{129,041^{1/1}}\right] = 297.7 \cong 298m$$

$$i_{\text{turbulent}} = \frac{l_{turbulent}}{L} = \frac{298}{396} = 0.752 \cong 75.2\%$$

(c) For the length of the transition zone, Eq. (2.46) is used:

$$l_{transition} = L \times \left[\frac{3,000^{1/\beta} - 2,000^{1/\beta}}{R_I^{1/\beta}}\right]$$

$$= 396 \times \left[\frac{3,000^1 - 2,000^1}{129,041^1}\right] = 3.1 \cong 3.0m$$

$$i_{transition} = \frac{l_{transition}}{L} = \frac{3}{396} = 0.0076 \cong 0.8\%$$

(d) Finally, the laminar flow length can be determined from Eq. (2.47):

$$l_{laminar} = L \times \left[\frac{2,000^{1/\beta} - 1}{R_I^{1/\beta}}\right]$$

$$= 396 \times \left[\frac{2,000^1 - 1}{129,041^1}\right] = 6.1 \cong 6m$$

$$i_{\text{laminar}} = \frac{l_{laminar}}{L} = \frac{6}{396} = 0.015 \cong 1.5\%$$

Case *(b)*: *Pressure head profile for improved nonuniform outflow concept* *[$\beta \neq 1, m_\beta = m \times \beta \neq (m = 1.828)$]:*

1. The nonuniform outflow exponent, β from Eq. (2.27):

First, the parameter, λ in the expression of β, is computed for m = 1.852: $\lambda = \frac{1-0.5^{m+1}}{m+2} = \frac{1-0.5^{2.828}}{3.828} = 0.224$.

For the previously known/computed values of parameters;

$H_{VI} = 0.084$ m, $H_{f(L)} = 2.843$ m, $\overline{H} = 31.61$ m, $x = 0.5$, p = 0 (horizontal pipe case, $\Delta S = S_0 \times L = 0$)

$$\beta = 1 - 3.322x \log\left[1 + \frac{H_{VI}}{\overline{H}} + \lambda\frac{H_{f(L)}}{\overline{H}} + \frac{1}{4\overline{H}}(p\Delta S)\right]$$

$$\beta = 1 - 3.322 \times 0.5$$
$$\times \log\left[1 + \frac{0.084}{31.61} + 0.224 \times \frac{2.843}{31.61} + \frac{1}{4 \times 31.61} \times (0)\right] = 0.9837 \cong 0.984$$

2. The improved value of velocity exponent, m_β:

$$m_\beta = m \times \beta = 1.828 \times 0.984 = 1.798$$

3. The Christiansen's friction correction factor, F_C for $\mu = s_I/s = 1.0$, from Eq. (2.16) [(a): Step 6]:

$$F_C = \frac{1}{(m_\beta + 1)} + \frac{1}{2N} + \frac{\sqrt{m_\beta - 1}}{6N^2}$$

$$F_C = \frac{1}{(1.798 + 1)} + \frac{1}{2 \times 33} + \frac{\sqrt{1.798 - 1}}{6 \times 33^2} = 0.3727$$

4. Total friction drop at the end of the pipeline, $H_{f(L)}$ [*(a): Step 8*]:

$$H_{f(L)} = F_C H_{f0} = 0.3727 \times 7.708 = 2.873 \text{ m}.$$

5. Operating inlet pressure head, H_{0I}, from Eq. (2.29) [*(a): Step 9*]:

$$H_{0I} = \overline{H} + \frac{(m_\beta + 1)}{(m_\beta + 2)} H_{f(L)} + p\frac{\Delta S}{2} = 31.61 + \frac{(1.798 + 1)}{(1.798 + 2)} \times 2.873 + 0 = 33.73m$$

6. The maximum pressure head, H_{max} from Eq. (2.30): [*(a): Step 10*]:

$$H_{max} = \overline{H} + H_{VI}(1 - \theta^2) + H_{f(L)}\left[\theta^{m_\beta+1} - \frac{1}{(m_\beta + 2)}\right] + p\Delta S\left(\theta - \frac{1}{2}\right)$$

$$\begin{aligned} H_{max} &= 31.61 + 0.084 \times (1 - 0.9697^2) \\ &\quad + 2.873 \times \left[0.9697^{(1.798+1)} - \frac{1}{(1.798 + 2)}\right] + 0 = 33.50 \text{ m} \end{aligned}$$

7. The minimum pressure head, H_{min}, which is equal to the downstream pressure head ($H_N = H_{min}$), from Eq. (2.21) [*(a): Step 11*]:

$$H_N = H_{min} = \overline{H} + H_{VI} - \left(\frac{1}{m_\beta + 2}\right) H_{f(L)} - \frac{p\Delta S}{2}$$

$$H_N = H_{min} = 31.61 + 0.084 - \frac{1}{(1.798 + 2)} \times 2.873 - 0 = 6.647 = 30.945m$$

8. The total friction drop for the pipe length (L_0) $H_{f(LO)}$, from the conservation of energy principle [*(a): Step 12*]:

$$H_{f(L_O)} \cong H_{max} + H_{VI} - H_N = 33.50 + 0.084 - 30.945 \cong 2.64m < H_{f(L)} = 2.843 \text{ m}.$$

9. The percentage of average pressure head, δ_H is evaluated by substituting Eq. (2.35) into Eq. (2.32) [*(a): Step 13*]:

$$\delta_H = \frac{\Delta H}{\overline{H}} = \frac{(H_{max} - H_{min})}{\overline{H}} \cong \theta^{m_\beta+1}\frac{H_{f(L)}}{\overline{H}} - \theta^2\frac{H_{VI}}{\overline{H}} + \theta\frac{(p\Delta S)}{\overline{H}}$$

$$\delta_H = \frac{H_{max} - H_{min}}{\overline{H}} = \frac{(33.50 - 30.945)}{31.61} = 0.08 = 8\% \quad \text{or}$$

$$\delta_H \cong 0.9697^{(1.798+1)} \times \frac{2.873}{31.61} - 0.9697^2 \times \frac{0.084}{31.61} + 0.9697 \times 0 = 0.08 = 8\%$$

10. The sprinkler flow variation, δ_q from Eq. (2.33) [*(a): Step 14*]:

$$\delta_q = 1 - (1 - \delta_H)^x = 1 - (1 - 0.08)^{0.5} = 0.041 \cong 4\%$$

11. The variance of pressure head profile H_{var}, can be evaluated from Eq. (2.37), for $m_\beta = 1.664$ [*(a): Step 15*]:

$$\begin{aligned} H_{var} = {} & \frac{8}{15} H_{VI}^2 + \frac{(m_\beta + 1)^2}{(m_\beta + 2)^2 (2m_\beta + 3)} H_{f(L)}^2 \\ & - \frac{4}{3} \frac{(m_\beta + 1)}{(m_\beta + 2)(m_\beta + 4)} H_{VI} H_{f(L)} - \frac{1}{6} H_{VI} (p\Delta S) \\ & + \frac{(m_\beta + 1)}{(m_\beta + 2)(m_\beta + 3)} H_{f(L)} (p\Delta S) + \frac{1}{12} (p\Delta S)^2 \end{aligned}$$

$$\begin{aligned} H_{var} = {} & \frac{8}{15} \times 0.084^2 + \frac{(1.798 + 1)^2}{(1.798 + 2)^2 \times (2 \times 1.798 + 3)} \times 2.873^2 \\ & - \frac{4}{3} \times \frac{(1.798 + 1)}{(1.798 + 2) \times (1.798 + 4)} \times 0.084 \times 2.873 - \frac{1}{6} \times 0.084 \times (0) \\ & + \frac{(1.798 + 1)}{(1.798 + 2) \times (1.798 + 3)} \times 2.873 \times (0) + \frac{1}{12} \times (0)^2 = 0.628 \cong 63\% \end{aligned}$$

12. The coefficient of variation of sprinkler outflow, CV_q from Eq. (2.38) [*(a): Step 16*]:

$$CV_q = \frac{x\sqrt{H_{var}}}{\overline{H}} = \frac{0.5 \times \sqrt{0.628}}{31.61} = 0.0125 = 1.25\%$$

Then, the coefficient of variation of pressure head, CV_H, from Eq. (2.38):

$$CV_H = \frac{CV_q}{x} = \frac{0.0125}{0.5} = 0.025 \cong 2.5\%$$

13. The Christiansen's uniformity coefficient, U_C from Eq. (2.39) [*(a): Step 17*]:

$$U_C = 1 - 0.798CV_q = 1 - 0.798 \times 0.0125 = 0.99 = 99\%$$

14. The lower-quarter distribution uniformity, DU_{LQ} is evaluted from the relationship between U_C and DU_{LQ} [Eq. (2.41)] or the relationship between DU_{LQ} and CV_q [Eq. (2.42)] [*(a): Step 18*]:

$$U_C = 0.37 + 0.63DU_{LQ}$$

$$DU_{LQ} = \frac{(U_C - 0.37)}{0.63} = \frac{(0.99 - 0.37)}{0.63} = 0.984 = 98.4\%$$

$$DU_{LQ} = 1 - 1.267CV_q = 1 - 1.267 \times 0.0125 = 0.984 = 98.4\%$$

15.

(a) The length of pipe inlet section in which fully turbulent flow occurs, is directly determined from Eq. (2.44):

$$l_{fully-turbulent} = L \times \left[1 - \left(\frac{10^5}{R_I}\right)^{1/\beta}\right]$$

$$= 396 \times \left[1 - \left(\frac{10^5}{129,041}\right)^{1/0.984}\right] = 90.4 \cong 90m$$

$$i_{fully-\text{turbulent}} = \frac{l_{fully-turbulent}}{L} = \frac{90}{396} = 0.228 \cong 23\%$$

(b) The turbulent flow length, $l_{\text{turbulent}}$ for the case of fully-turbulent flow exists at the inlet of the pipeline, can be determined Eq. (2.45a):

$$l_{turbulent} = L \times \left[\frac{10^{5/\beta} - 3000^{1/\beta}}{R_I^{1/\beta}}\right]$$

$$= 396 \times \left[\frac{10^{5/0.984} - 3,000^{1/0.984}}{129,041^{1/0.984}} \right] = 296.95 \cong 297m$$

$$i_{\text{turbulent}} = \frac{l_{turbulent}}{L} = \frac{297}{396} = 0.7498 \cong 75\%$$

(c) For the length of the transition zone, Eq. (2.46) is used:

$$l_{transition} = L \times \left[\frac{3,000^{1/\beta} - 2,000^{1/\beta}}{R_I^{1/\beta}} \right]$$

$$= 396 \times \left[\frac{3,000^{1/0.984} - 2,000^{1/0.984}}{129,041^{1/0.984}} \right] = 2.925 \cong 3.0m$$

$$i_{transition} = \frac{l_{transition}}{L} = \frac{3}{396} = 0.0076 \cong 0.8\%$$

(d) Finally, the laminar flow length can be determined from Eq. (2.47):

$$l_{laminar} = L \times \left[\frac{2,000^{1/\beta} - 1}{R_I^{1/\beta}} \right]$$

$$= 396 \times \left[\frac{2,000^{1/0.984} - 1}{129,041^{1/0.984}} \right] = 5.73 \cong 6m$$

$$i_{laminar} = \frac{l_{laminar}}{L} = \frac{6}{396} = 0.015 \cong 1.5\%$$

For the sake of comparison of the complete results obtained from the R-EGL method for the design cases (a): uniform outflow profile, and (b): nonuniform outflow profile, are clearly synthesized in Table 2.2.

Comparison Test with Step-By-Step (SBS) Procedure

To prove degree of accuracy of the present R-EGL method, for the present design examples (Examples **1** and **2**) the results of the analytical solution are compared with those obtained from the computer-aided step-by-step (SBS) method [7, 8]. The solution of the SBS method was obtained using by a computer program developed in *Visual Basic 6.0* [40], based on the flowcharts presented in the existent literature [7, 8]. Moreover, the computer program was comprehensively examined for various design combinations and different line slope cases with respect to the large ranges of the inlet pressure head, the pipe diameter and the pipe length [34, 35].

Table 2.2 Comparison of results for hydraulic variables computed from the R-EGL Method using input parameters given in Design Example 2 (sprinkler lateral design) for **(a)**: uniform outflow profile ($\beta = 1$) and **(b)**: nonuniform outflow profile ($\beta = 0.984$)

	Results Obtained from the R-EGL Method for:																		
	Pressure Parameters:						Uniformity Parameters:							Flow Regime Parameters:					
	F_C	$H_{f(L)}$	$H_{f(Lo)}$	H_{0I}	H_{max}	$H_{min} = H_N$	δ_H	δ_q	H_{var}	CV_q	CV_H	U_C	DU_{LQ}	$l_{fully\text{-}turbulent}$	$l_{turbulent}$	$l_{laminar}$	$i_{fully\text{-}turb}$	$i_{turbulent}$	$i_{laminar}$
	(—)	(m)	(m)	(m)	(m)	(m)	(%)	(%)	(%)	(%)	(%)	(%)	(%)	(m)	(m)	(m)	(%)	(%)	(%)
Equation number:	*(4.16)*	*(4.10)*	*(4.10)*	*(4.21)* *(4.29)*	*(4.30)*	*(4.21)*	*(4.32)* *(4.35)*	*(4.33)*	*(4.37)*	*(4.38)*	*(4.38)*	*(4.39)* *(4.40)*	*(4.41)* *(4.42)*	*(4.44)*	*(4.45a)*	*(4.47)*	*(—)*	*(—)*	*(—)*
(a): Uniform Outflow Profile	0.3689	2.843	2.61	33.71	33.48	30.96	8	4	62	1.25	2.5	99	98.4	89	298	6	23	75	2
(b): Nonuniform Outflow Profile	0.3727	2.873	2.64	33.73	33.50	30.94	8	4	63	1.25	2.5	99	98.4	90	297	6	23	75	2

(a): Uniform Outflow Profile ($\beta = 1$, m = 1.828) and **(b): Nonuniform** Outflow Profile ($\beta = 0.984$, $m_\beta = 1.798$)

$\overline{H} = 31.61$ m, $\theta = L_0/L = 0.9697$, $x = 0.5$, p = 0 (Horizontal pipe case), $Q_I = 1.04 \times 10^{-2}$ m^3s^{-1}, $H_{VI} = 0.084$ m, $R_I = 129{,}041$, $H_{f0} = 7.71$ m,

Fully-turbulent flow situation ($10^5 < R_I = 129{,}041 < 10^5$), $f = 0.0218$ (relative roughness, $\varepsilon/D = 0.1/101.6 = 9.84 \times 10^{-4}$)

Comparative Analysis for Design Example: 1

In order to do a detailed comparative analysis, the solution of the Design Example **1** is extended to cover various combinations of design parameters, varying emitter discharge exponents (x = 0.2, 0.5, 0.54 and 1.0) and different uniform line slope conditions (S_0 = 0.00, −0.02, −0.05, 0.02 and, 0.05) and, the complete results were presented in Table 2.3. As shown in this table, for the Design Example **1** (S_o = 0.0 and x = 0.54), for the pressure heads at the inlet, H_I, at the first and last emitters, H_1, and H_N; the analytical solution yields 8.70 m, 8.66 m and 6.66 m respectively, whereas the SBS numerical method yields the same value for both H_I and H_1 as 8.86 m and, 6.91 m for H_N. Note that, the H_I is equivalent to H_1 disregarding inlet spacing from the main pipe to the first emitter, for the SBS trickle lateral design procedure [7].

As shown in this table, for all design cases the results obtained from the analytical solution and those of the SBS numerical method are in a good justification. For the friction head loss for length of the pipe section L0 between the first and last emitters $H_{f(Lo)}$ and, the total friction head losses for the entire length of the pipeline L including inlet spacing to the first emitter $H_{f(L)}$; the proposed method gives 2.04 m and 2.08 m respectively, whereas the SBS method yields for $H_{f(Lo)}$ 1.93 m. For the uniformity design parameters δ_q, CV_q and U_C; the proposed method gives 13.4%, 4.5% and 96.4% whereas the SBS method yields nearly similar values 12.5%, 4.1% and 96.7%, respectively.

The performance of the proposed analytical method for all design combinations examined is satisfactory especially for a large range of variation in the uniformity coefficient U_C. According to the both analytical and numerical methods, for a downhill slope, S_o = −0.02, and for different type of emitters, the design uniformity parameters δ_q, CV_q and U_C yield higher values than those of the other design slope combinations. For the large values of uphill and downhill uniform slope ranges (S_o = 0.05 and S_o = −0.05), for the uniformity design criteria, the both methods tend to give smaller values than those of the other design slope combinations. Table 2.3 indicates, for all design slope combinations, the design uniformity parameters (δ_q, CV_q and U_C) give smaller values in accordance with increasing in the emitter discharge exponent; for instances, for laminar flow emitters (x = 1.0) these parameters have the smallest values and, for the turbulent flow emitters (x = 0.5 and 0.54) these parameters have values between those of the laminar flow emitter and the emitter discharge exponent, x = 0.2. Flow characteristics for various design combinations computed from the both methods can be comparable from Table 2.3.

For the zero slope case (S_0 = 0) and, for the turbulent flow emitter (x = 0.54), the flow characteristics versus the dimensionless distance i = l/L, obtained from both the procedures are also evaluated in Figs. 2.4 and 2.5, respectively. Figure 2.4 shows variation of the pressure head H(l), the outflow q(l), the flow rate Q(l),and the Reynolds number R(l) along the lateral line, whereas the profile of the friction drop ratio $R_i = H_f(l)/H_{f(L)}$ [Eq. (2.9)], is shown in Fig. 2.5.

Table 2.3 Comparison of hydraulic design parameters for trickle lateral design based on the proposed analytical method and the numerical Step-By-Step (SBS) method for various combinations of uniform line slope cases ($S_o = 0.0, -0.02, -0.05, 0.02$ and 0.05) and, different types of emitter ($x = 0.2, 0.5, 0.54$ and 1.0) using given data in Design Example 1

Slope (S_O)	Present Analytical Revised Energy-Grade Line (REGL) Method												Computer-Aided Step-by-Step (SBS) Method								
	x	H_I (m)	H_1 (m)	H_N (m)	H_{max} (m)	H_{min} (m)	N_{min}	$H_{f(Lo)}$ (m)	$H_{f(L)}$ (m)	δ_q (%)	CV_q (%)	U_C (%)	$H_I = H_1$ (m)	H_N (m)	H_{max} (m)	H_{min} (m)	N_{min}	$H_{f(Lo)}$ (m)	δ_q (%)	CV_q (%)	U_C (%)
0.00	0.2	8.70	8.66	6.66	8.66	6.66	151	2.02	2.05	5.1	1.6	98.7	8.86	6.91	8.86	6.91	151	1.97	4.8	1.5	98.8
	0.5	8.72	8.68	6.64	8.68	6.64	151	2.06	2.09	12.5	4.1	96.7	8.70	6.78	8.70	6.78	151	1.94	11.7	3.9	96.9
	0.54	8.71	8.67	6.65	8.67	6.65	151	2.04	2.08	13.4	4.5	96.4	8.70	6.79	8.70	6.79	151	1.93	12.5	4.1	96.7
	1.0	8.76	8.72	6.61	8.72	6.61	151	2.12	2.16	24.2	8.6	93.1	8.57	6.69	8.57	6.69	151	1.89	21.9	7.9	93.7
−0.02	0.2	7.17	7.15	8.19	8.19	6.82	44	1.98	2.02	3.6	1.1	99.1	7.14	8.12	8.12	6.75	44	2.04	3.6	1.1	99.1
	0.5	7.16	7.14	8.20	8.20	6.81	43	1.96	1.99	8.8	2.9	97.7	7.26	8.10	8.10	6.93	43	2.17	7.5	2.9	97.7
	0.54	7.16	7.14	8.20	8.20	6.77	43	1.96	1.99	9.8	3.0	97.6	7.25	8.23	8.23	6.91	43	2.03	9.0	2.9	97.7
	1.0	7.14	7.12	8.22	8.22	6.75	42	1.92	1.96	17.9	5.7	95.4	7.24	8.04	8.04	6.87	44	2.22	14.5	5.7	95.4
−0.05	0.2	4.87	4.89	10.48	10.48	4.89	1	1.92	1.95	14.1	4.6	96.3	5.17	10.66	10.66	5.17	1	2.03	13.5	5.0	96.0
	0.5	4.81	4.83	10.53	10.53	4.83	1	1.81	1.85	32.3	11.9	90.5	5.11	10.37	10.37	5.11	1	2.25	29.8	12.9	89.7
	0.54	4.81	4.82	10.54	10.54	4.82	1	1.79	1.83	34.4	12.9	89.7	5.10	10.37	10.37	5.10	1	2.23	31.8	12.9	89.7
	1.0	4.73	4.74	10.59	10.59	4.74	1	1.66	1.69	55.2	24.5	80.4	5.02	10.54	10.54	5.02	1	1.99	52.4	23.8	81.0
0.02	0.2	10.23	10.17	5.14	10.17	5.14	151	2.05	2.09	12.7	4.0	96.8	10.44	5.56	10.44	5.56	151	1.90	11.8	4.0	96.8
	0.5	10.28	10.22	5.09	10.22	5.09	151	2.15	2.19	29.4	10.3	91.8	10.21	5.39	10.21	5.39	151	1.83	27.3	10.0	92.0
	0.54	10.29	10.23	5.08	10.23	5.08	151	2.17	2.20	31.5	11.1	91.1	10.15	5.35	10.15	5.35	151	1.82	29.2	10.8	91.4
	1.0	10.37	10.32	4.99	10.32	4.99	151	2.34	2.37	51.6	21.4	82.9	10.62	5.33	10.62	5.33	151	2.31	49.8	19.7	84.3
0.05	0.2	12.52	12.44	2.85	12.44	2.85	151	2.11	2.14	25.5	7.6	93.9	12.89	3.44	12.89	3.44	151	1.97	23.2	7.9	93.7

(continued)

Table 2.3 (continued)

Slope (S_0)	Present Analytical Revised Energy-Grade Line (REGL) Method										Computer-Aided Step-by-Step (SBS) Method										
	x	H_I (m)	H_1 (m)	H_N (m)	H_{max} (m)	H_{min} (m)	N_{min}	$H_{f(Lo)}$ (m)	$H_{f(L)}$ (m)	δ_q (%)	CV_q (%)	U_C (%)	$H_I = H_1$ (m)	H_N (m)	H_{max} (m)	H_{min} (m)	N_{min}	$H_{f(Lo)}$ (m)	δ_q (%)	CV_q (%)	U_C (%)
	0.5	12.62	12.53	2.75	12.53	2.75	151	2.29	2.33	53.1	19.7	84.3	12.51	3.36	12.51	3.36	151	1.66	48.2	19.7	84.3
	0.54	12.63	12.55	2.74	12.55	2.74	151	2.32	2.36	56.0	21.3	83.0	12.40	3.26	12.40	3.26	151	1.66	51.4	21.2	83.1
	1.0	12.81	12.72	2.54	12.72	2.54	151	2.70	2.74	80.0	41.3	67.1	12.94	3.08	12.94	3.08	151	2.38	76.2	38.1	69.6

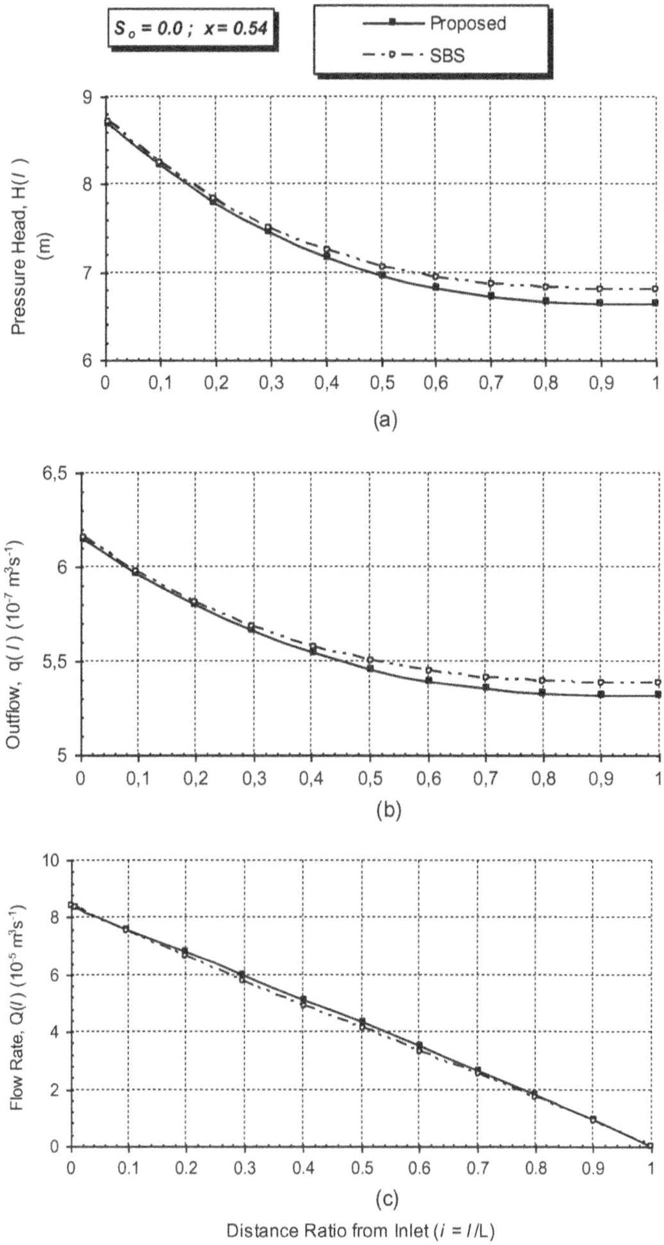

Fig. 2.4 **a** Pressure head, **b** outflow, and **c** flow rate profiles for horizontal trickle lateral with respect to the distance ratio from inlet, for the emitter discharge exponent (x = 0.54), obtained from the presented analytical solution and the numerical SBS solution using given data in Design Example 1

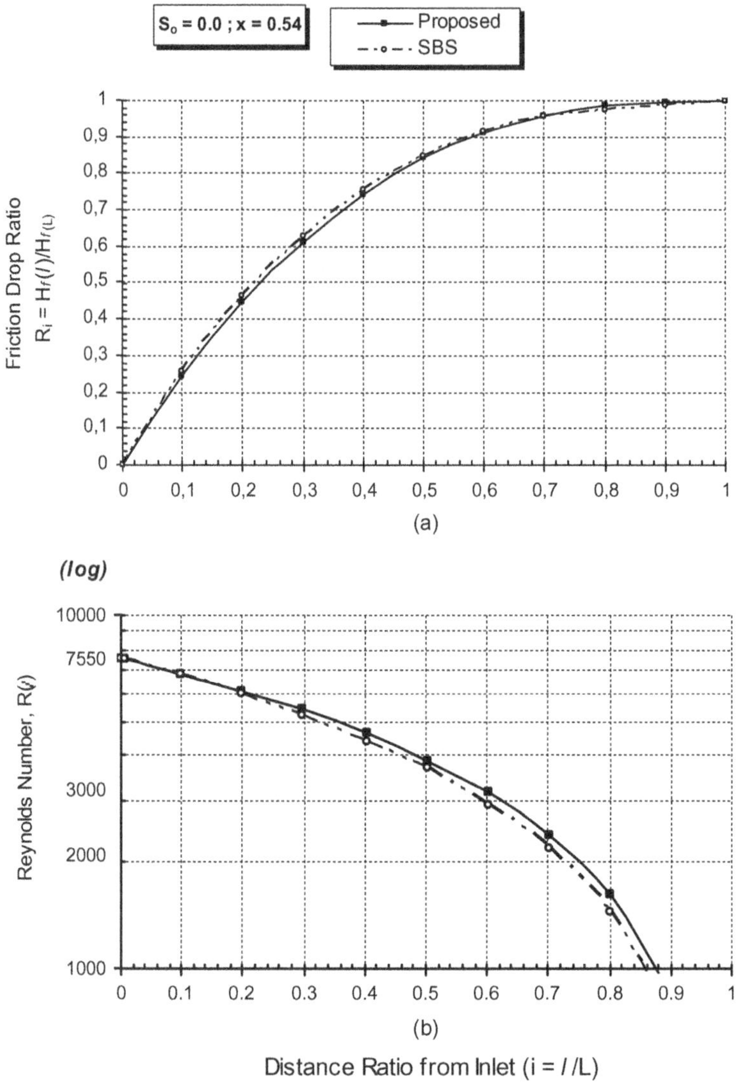

Fig. 2.5 Distribution of **a** friction drop and **b** Reynolds number for horizontal trickle lateral with respect to the distance ratio from inlet, for the emitter discharge exponent ($x = 0.54$), obtained from the presented analytical solution and the numerical SBS solution using given data in Design Example 1

Figure 2.4 illustrates in the middle point of the lateral (at the middle emitter) total reduction (in comparison with the initial values at the lateral inlet) in emitter pressure head and outflow are about 80% and 89% (which are approximately equal to the average values; $\overline{H}, \overline{q}$) and, in lateral flow rate is about 52% are occurred, respectively, according to the both methods. The flow rate values Q(l), from both the methods are in a good agreement in each lateral section; mean and maximum deviation from the accurate SBS method is about 3% and 5%, respectively (Fig. 2.4c).

As shown in Fig. 2.5a, for the both methods, about 84% of the total friction drop occurs at the middle emitter (i = 50%), whereas half of the total friction drop occurs upstream of emitter No. 35 (i = 23%) and, about 75% of the total friction drop occurs upstream of emitter No. 60 (i = 40%). It is obvious that, the friction drop ratio in the upstream section is more than that of downstream section of a multiple outlets pipeline.

As shown in Fig. 2.5b, different flow regime zones (turbulent, transition and, laminar) from the inlet section and, their lengths across the length of trickle lateral, were also presented. From this figure, according to both the methods, between l = 113 m and 114 m ($i_{turbulent}$ = 75%) flow becomes laminar (including the transition zone) and, for the downstream length, l = 38 m, flow is actually laminar ($i_{laminar}$ = 25%).

Figure 2.6 shows according to the both methods, for different downhill and uphill slope ranges ($S_o = -0.02, -0.05, 0.02$ and 0.05) and for the value of emitter discharge exponent x = 0.54, the pressure head H(l) profiles versus the dimensionless distance ratio, $i = l/L$. As shown in Fig. 2.6a, for a downhill slope, $S_o = -0.02$, the global maximum operating emitter pressure head, H_{max}, occurs at the first emitter near the downstream closed end whereas the minimum operating emitter pressure head, H_{min}, occurs for the emitter in the upstream segment, N_{min} = 43 (i = 0.28), for the both methods. The analytical solution yields for H_{max} and H_{min}, 8.20 m and 6.77 m respectively, whereas the SBS numerical solution yields the similar values 8.23 m and 6.91 m, respectively (Table 2.3).

In the same figure (Fig. 2.6), for an extreme value of the downhill slope, $S_o = -0.05$, the global maximum operating emitter pressure head, H_{max}, occurs at the first emitter near the downstream closed end whereas the minimum operating emitter pressure head, H_{min}, occurs at the first emitter near the lateral inlet, according to the both methods.

Comparative Analysis for Design Example: 2

For the sake of comparison, for the extension of the Design Example **2**, hydraulic design parameters were also computed for zero slope case ($S_0 = 0$) and two different values of lateral diameters (D = 76.2 mm and 101.6 mm) and, the complete results are presented in Table 2.4. As shown in this table, for the value of pipe diameter D = 101.6 mm, for the pressure heads at the inlet, first and last sprinklers, H_I, H_1, and H_N; the proposed analytical solution yields 33.72 m, 33.50 m and 30.94 m, whereas the SBS numerical method yields the similar values 33.80 m, 33.60 m and 30.88 m, respectively.

For the friction head loss for length of the pipe section L0 between the first and last sprinklers $H_{f(Lo)}$ and, the total friction head losses for the entire length of the pipeline

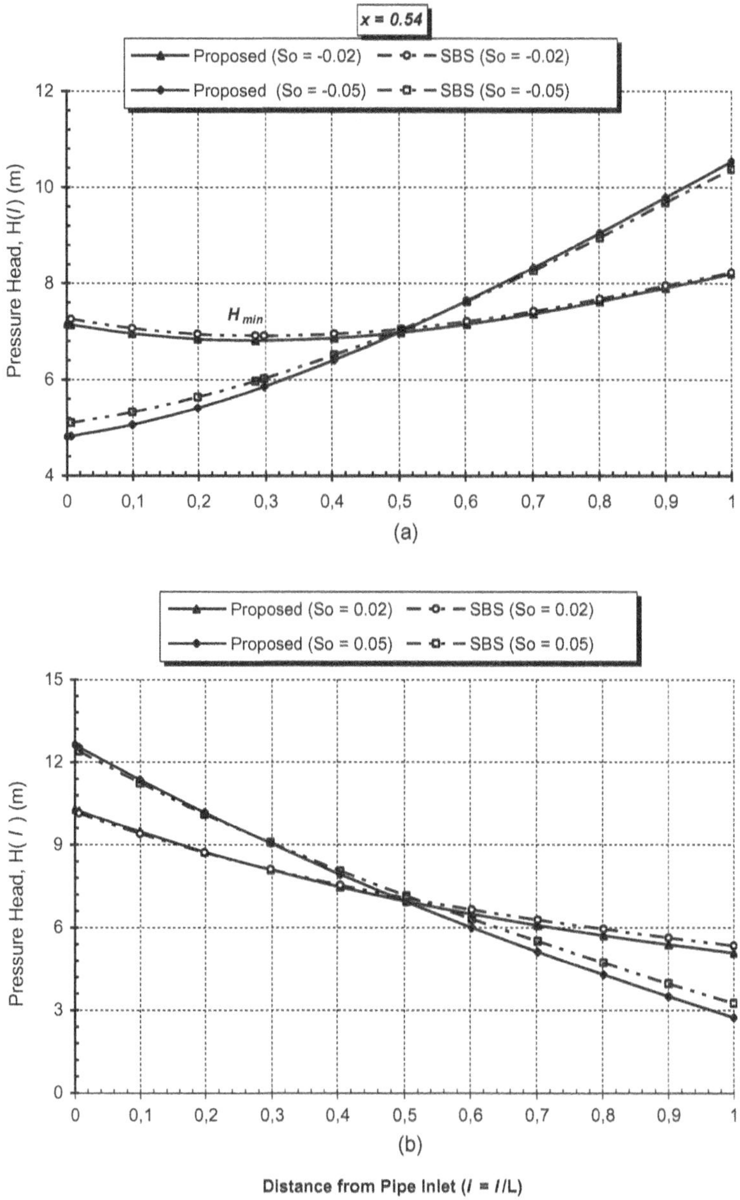

Fig. 2.6 Pressure head profile for trickle lateral design with respect to the distance ratio from inlet, for the uniform (a): downhill ($S_O = -0.02$ and -0.05), and (b): uphill ($S_O = 0.02$ and 0.05) line slope cases, and for the emitter discharge exponent ($x = 0.54$), obtained from the presented analytical solution and the numerical SBS solution using given data in Design Example 1

Table 2.4 Comparison of hydraulic design parameters for horizontal sprinkler lateral design based on the proposed analytical method and the numerical Step-By-Step (SBS) method for two lateral diameter values (D = 76.2 mm and 101.6 mm) using given data in Design Example 2

Internal Diameter D (mm)	Proposed Analytical Method								Numerical Step-by-Step Method (SBS)							
	H_I (m)	$H_1 = H_{max}$ (m)	$H_N = H_{min}$ (m)	$H_{f(Lo)}$[a] (m)	$H_{f(L)}$[b] (m)	δ_H (%)	CV_q (%)	U_C (%)	H_I (m)	$H_1 = H_{max}$ (m)	$H_N = H_{min}$ (m)	$H_{f(Lo)}$[a] (m)	$H_{f(L)}$[b] (m)	δ_H (%)	CV_q (%)	U_C (%)
76.2 (3 in.)	41.01	39.90	28.40	11.76	12.87	36.4	5.76	95.4	40.90	39.90	28.80	11.36	12.36	35.1	5.26	95.8
101.6 (4 in.)	33.72	33.50	30.94	2.64	2.87	8.1	1.26	98.9	33.80	33.60	30.88	2.80	3.0	8.6	1.26	98.9

[a] Friction head loss for length of the pipe section L_o, between the first and last sprinklers;

[b] Total friction head losses for the entire length of the pipeline L, including inlet spacing to the first sprinkler

L including inlet spacing to the first sprinkler $H_{f(L)}$; the analytical solution gives 2.64 m and 2.87 m respectively, whereas the SBS method gives bigger values 2.80 m and 3.0 m, respectively. According to the both methods, about 86% of the total friction drop occurs at the middle sprinkler ($i = 50\%$), whereas half of the total friction drop occurs upstream of sprinkler No. 7 ($i = 22\%$) and, about 75% of the total friction drop occurs upstream of sprinkler No. 13 ($i = 38\%$). For the uniformity design parameters δ_H, CV_q and U_C; the proposed method yields 8.1%, 1.26% and 98.9% whereas the SBS method yields nearly the same values 8.6%, 1.26% and 98.9%, respectively.

For the sake of comparison from Table 2.4, the results obtained from the analytical solution and the SBS numerical method for sprinkler lateral design are also in a good justification for the both lateral diameters, D = 76.2 and 101.6 mm. As a matter of fact, maximum relative error for pressure heads and uniformity levels is less than 0.5%, for the both design cases examined (Table 2.4).

Concluding Remarks

In this chapter, an analytical procedure that is simple, direct and efficient to solve hydraulic design problems of various types of multiple outlets pipelines, was developed. A comprehensive comparison test applied for various design combinations and different line slope cases indicates that, the results of the proposed analytical procedure had been found to be sufficiently accurate in comparison with the accurate numerical stepwise (SBS) method. The performance of this procedure for all design cases examined is satisfactory especially for a large range of variation in the uniformity coefficient U_C. The analytical development presented is preferred to the numerical techniques which need large amounts of execution time and complex computer operations.

Based on the presented R-EGL method, the following conclusions are underlined:

1. The Christiansen's friction correction factor (F_C) accurately predicts the friction head loss since F_C is an accurate approximation of the stepwise correction factor (F_{sw}). The traditional correction factor (F) tends to underpredict systematically with respect to the F_C and F_{SW} for finite number of outlets; particularly, as the number of outlets have the extreme small values, the deviation from these factors (F_C and F_{SW}) being more significant.
2. The friction drop ratio [$R_i = H_f(l)/H_{f(L)}$] is an important indicator to show the friction drop profile along a multiple outlets pipeline. Investigation of the friction head loss distribution along the laterals for both trickle and sprinkler irrigation system shows that, for zero slope case, at the middle outlet (emitter or sprinkler) about 85% of the total friction drop occurs, whereas half of the total friction drop occurs at about 22% of the full length of the lateral and, 75% of the total friction drop occurs at about 39% of the full length of the lateral, respectively. As a consequence, the friction drop ratio

in the upstream section is more than that of downstream section of a multiple outlets pipeline.

3. The average head loss due to friction for the entire length of the lateral ($\overline{H}_f$) and, for the upper half length of the lateral ($\overline{H}_{f0.5}$) are about 73% and 51% of the total friction drop at the downstream closed end of the lateral ($H_{f(L)}$), respectively, when the Darcy-Weisbach formula is used.
4. For trickle lateral design in zero slope case, in the middle of the lateral, total reduction (with respect to the initial values at the lateral inlet) in emitter pressure head and outflow are about 80% and 89% (which are approximately equal to the average values; $\overline{H}, \overline{q}$) and, in lateral flow rate is about 52% are occurred, respectively.
5. The pressure head profile for the nonuniform outflow concept indicates, the value of the pressure head (or outflow) at the middle outlet is independent from the various uphill and downhill uniform line slope conditions and, its value at this point is approximately equal to the average pressure head (or outflow).
6. In various design combinations of trickle lateral to cover varying emitter discharge exponents and, different uniform line slope conditions; for the downhill slope case, $S_o = -0.02$, for the design uniformity parameters (δ_q, CV_q and U_C), both the proposed analytical method and SBS numerical method yield higher values than those of the other slope combinations whereas for large uphill and downhill uniform slope cases ($S_o = 0.05$ and $S_o = -0.05$), both the methods tend to give smaller values than those of the other slope combinations.
7. For trickle lateral design, for all design slope combinations, the design uniformity parameters (δ_q, CV_q and U_C), give smaller values with increasing in the emitter discharge exponent x; for instances for a value of $x = 0.2$, the uniformity parameters have the highest values, whereas for the laminar flow emitters ($x = 1.0$) they have the smallest values.

Acknowledgements The senior author wants to express his acknowledges to Prof. Vijay. P. Singh at Texas A&M University, Emeritus Prof. Necati Ağıralioğlu from International Antalya Science University, and Distinguished Emeritus Professor Zekai Şen from Istanbul Medipol University, for their highly encouragement on preparing this book, many scientific activities, projects and so on. The Scientific and Technological Research Council of Turkey (TUBITAK) is acknowledged for supporting by the fellowships and grant program during the studies in the Texas A&M University/BAEN. We also thank to publishers, potential readers and reviewers for providing constructive reviews and criticism on the content of the book, in advance.

Conflict of Interest The author/editor has no conflicts of interest, financial or otherwise, to declare.

Appendix: Relationship Between Coefficient of Variation of Discharge (CVq) and Christiansen's Uniformity Coefficient (U_C)

1. The variance of pressure head profile between the pipe segments $l = 0$ and $l = L$, can be expressed as [1]

$$H_{\text{var}} = \frac{1}{L}\int_{l=0}^{l=L} DEV_l^2 dl = \frac{1}{L}\int_{l=0}^{l=L} \left[H(l) - \overline{H}\right]^2 dl \tag{2.48}$$

where DEV_l = deviation between $H(l)$ and $\overline{H}$ at any distance l from the pipeline inlet.
Subtracting $\overline{H}$ from Eq. (2.21) and into Eq. (2.48) then, solving the integration yields

$$\begin{aligned} H_{\text{var}} = {} & \frac{8}{15}H_{VI}^2 + \frac{(m_\beta+1)^2}{(m_\beta+2)^2(2m_\beta+3)}H_{f(L)}^2 - \frac{4}{3}\frac{(m_\beta+1)}{(m_\beta+2)(m_\beta+4)}H_{VI}H_{f(L)} \\ & - \frac{1}{6}H_{VI}(p\Delta S) + \frac{(m_\beta+1)}{(m_\beta+2)(m_\beta+3)}H_{f(L)}(p\Delta S) + \frac{1}{12}(p\Delta S)^2 \end{aligned} \tag{2.49}$$

2. The relationship between CV_q and CV_H is given by [3]

$$CV_q = xCV_H = \frac{x\sqrt{H_{\text{var}}}}{\overline{H}} \tag{2.50}$$

in which CV_q = the coefficient of variation of discharge is statistically defined as the standard deviation of discharge divided by the mean value and expressed in percent; and CV_H = the coefficient of variation of pressure head defined as standard deviation of pressure head divided by the average pressure head and expressed in percent. The commonly used level of the CV_q varies for the intervals 3% and 5% [1, 26, 41].

3. For normal distribution, the Christiansen's uniformity coefficient U_C is related to CV_q by [38]

$$U_C = 1 - 0.798CV_q \tag{2.51}$$

4. Substituting Eq. (2.49) for H_{var} into Eq. (2.50), then the CV_q into Eq. (2.51); finally, the U_C coefficient can be directly evaluated from Eq. (2.40), presented in the text.

References

1. J. Keller, R.D. Bliesner, *Sprinkler and Trickle Irrigation* (Van Nostrand Reinhold, New York, 1990)
2. D.S. Miller, *Internal Flow Systems* (BHRA Publications, Cranfield, UK, 1990), pp. 189–203
3. V. Ravikumar, C.R. Ranganathan, S. Santhana Bosu, Analytical equation for variation of discharge in drip irrigation laterals. J. Irrig. and Drain. Eng., ASCE **129**(4), 295–298, 2003
4. E.J. Scaloppi, R.G. Allen, Hydraulics of irrigation laterals: comparative analysis. J. Irrig. and Drain. Eng., ASCE **119**(1), 91–115 (1993)
5. J.D. Valiantzas, Analytical approach for direct drip lateral hydraulic calculation. J. Irrig. and Drain. Eng., ASCE **124**(6), 300–305 (1998)
6. J.D. Valiantzas, Continuous outflow variation along irrigation laterals: effect of the number of outlets. J. Irrig. and Drain. Eng., ASCE **128**(1), 34–42 (2002)
7. P. Vallesquino, P.L. Luque-Escamilla, New algorithm for hydraulic calculation in irrigation laterals. J. Irrig. and Drain. Eng., ASCE **127**(4), 254–260 (2001)
8. P. Vallesquino, P.L. Luque-Escamilla, Equivalent friction factor method for hydraulic calculation in irrigation laterals. J. Irrig. and Drain. Eng., ASCE **128**(5), 278–286 (2002)
9. R.D. von Bernuth, Simple and accurate friction loss equation for plastic pipe. J. Irrig. and Drain. Eng., ASCE **116**(2), 294–298 (1990)
10. R.D. von Bernuth, T. Wilson, Friction factors for small diameter plastic pipes. J. Hydraulic Eng., ASCE **115**(2), 183–192 (1989)
11. A.W. Warrick, M. Yitayew, Trickle lateral hydraulics.I: analytical solution. J. Irrig. and Drain. Eng., ASCE **114**(2), 281–288 (1988)
12. G.Z. Watters, J. Keller, Trickle irrigation tubing hydraulics. Presented at the *Summer Meeting of ASAE*, Utah State University, Logan, Utah, Paper No. 78-2015 (1978)
13. I.P. Wu, Energy gradient line approach for direct hydraulic calculation in drip irrigation design. Irrig. Sci. **13**, 21–29 (1992)
14. I.P. Wu, An assessment of hydraulic design of microirrigation systems. Agric. Water Manage., 32, 275-284 (1997)
15. I.P. Wu, H.M. Gitlin, Hydraulics and uniformity for drip irrigation. J. Irrig. and Drain. Div., ASCE, 99(IR2), Proc. Paper 9786, June, pp. 157–168 (1973)
16. I.P. Wu, H.M. Gitlin, Drip irrigation design based on uniformity. Trans. ASAE **17**(3), 429–432 (1974)
17. I.P. Wu, H.M. Gitlin, Energy gradient line for drip irrigation laterals. J. Irrig. and Drain. Eng., ASCE 101(4), 323–326, 1975.
18. I.P. Wu, R. Yue, Drip lateral design using energy gradient line approach. Trans. ASAE **36**(2), 389–394 (1993)
19. G. Yıldırım, "Analytical relationships for designing multiple outlets pipelines. " J. Irrig. and Drain. Eng. ASCE **133**(2), 140–154 (2007)
20. H.M. Hathoot, A.I. Al-Amoud, F.S. Mohammad, Analysis and design of trickle irrigation laterals." J. Irrig. and Drain. Eng., ASCE **119**(5), 756–767 (1993)

21. H.M. Hathoot, A.I. Al-Amoud, F.S. Mohammad, Analysis and design of sprinkler irrigation laterals. J. Irrig. and Drain. Eng., ASCE **120**(3), 534–549 (1994)
22. Y. Kang, S. Nishiyama, Analysis and design of microirrigation laterals. J. Irrig. and Drain. Eng., ASCE **122**(2), 75–82 (1996)
23. Y. Kang, S. Nishiyama, Design of microirrigation submain unit. J. Irrig. and Drain. Eng., ASCE **122**(2), 83–89 (1996)
24. G. Yıldırım, N. Ağıralioğlu, Variation of total friction head losses and uniformity coefficients based on inlet pressure head in trickle laterals,in *Proc., International Conference on Water Resources Management in Arid Regions* (Kuwait Institute for Scientific Research, Kuwait, 2002)
25. G. Yıldırım, N. Ağıralioğlu, Discussion of 'Continuous outflow variation along irrigation laterals: effect of the number of outlets' by J. D. Valiantzas. J. Irrig. and Drain. Eng., ASCE **129**(5), 382–386 (2003)
26. G. Yıldırım, N. Ağıralioğlu, Linear solution for hydraulic analysis of tapered microirrigation laterals. J. Irrig. and Drain. Eng., ASCE **130**(1), 78–87 (2004)
27. G. Yıldırım, N. Ağıralioğlu, Comparative analysis of hydraulic calculation methods in design of microirrigation laterals. J. Irrig. and Drain. Eng., ASCE **130**(3), 201–217 (2004)
28. G. Yıldırım, N. Ağıralioğlu, Variation of design parameters in microirrigation laterals. ARI Interdiscip J Phys Eng Sci, 54(1), 42–53 (2004)
29. G. Yıldırım, and N. Ağıralioğlu, Discussion of 'Inlet pressure, energy cost, and economic design of tapered irrigation submains' by J. D. Valiantzas. J. Irrig. and Drain. Eng., ASCE **131**(2), 220–224 (2005)
30. G. Yıldırım, N. Ağıralioğlu, Closure to 'Linear solution for hydraulic analysis of tapered microirrigation laterals' by G. Yıldırım and N. Ağıralioğlu. J. Irrig. and Drain. Eng., ASCE **131**(5), 490–491 (2005)
31. I.P. Wu, Personal communication (2002)
32. G. Yıldırım, O. Kose, New Software (Multi-flowCAD) Development for Sub-main Line Hydraulic Design: Computational Algorithm, Computer Visualization and Implementation. Arab. J. Sci. Eng. **41**, 1551–1560 (2016)
33. G.Yıldırım, Computer-based analysis of hydraulic design variables for uniformly sloping microirrigation system laterals. J. Irrig. and Drain. Eng., ASCE **141**(7), 04014079 (2015)
34. G. Yıldırım, An Assessment for Hydraulically Efficient Design of Uniformly Sloping Submain, Lines. Irrig. Drain. **64**(2), 157–168 (2015)
35. G. Yıldırım, Total Energy Loss Assessment for Trickle Lateral Lines Equipped with Integrated In-Line and On-Line Emitters. Irrig. Sci. **28**(4), 341–352 (2010)
36. G. Yıldırım, V.P. Singh, A *MathCAD* Procedure for Commercial Pipeline Hydraulic Design Considering Local Energy Losses. Adv. Eng. Softw. **41**(2), 489–496 (2010)
37. G. Yıldırım, Computer-Based Analysis of Explicit Approximations to the Implicit Colebrook-White Equation in Turbulent Flow Friction Factor Calculation. Adv. Eng. Softw. **40**(11), 1183–1190 (2009)
38. G. Yıldırım, M. Özger, Neuro-Fuzzy Approach in Estimating Hazen-Williams Friction Coefficient for Small-Diameter Polyethylene Pipes. Adv. Eng. Softw. **40**(8), 593–599 (2009)
39. G. Yıldırım, Simplified Procedure for Hydraulic Design of Small-Diameter Plastic Pipes. Irrig. and Drain. (ICID Journal) **58**(3), 209–233 (2009)
40. M. Özger, G. Yıldırım, Determining turbulent flow friction coefficient using adaptive neuro-fuzzy computing technique. Adv Eng Softw **40**(4): 281–287 (2009)
41. G. Yıldırım, Determining operating inlet pressure head incorporating uniformity parameters for multioutlet plastic pipelines. J. Irrig. and Drain. Eng., ASCE **134**(3), 341–348 (2008)
42. G. Yıldırım, An assessment of hydraulic design of trickle laterals considering effect of minor losses. Irrig. and Drain. ICID J. **56**(4), 399–421 (2007)

43. G. Yıldırım, Hydraulic analysis and direct design of multiple outlets pipelines laid on flat and sloping lands. J. Irrig. and Drain. Eng., ASCE **132**(6), 537–552 (2006)
44. G. Yıldırım, Discussion of "Statistical Analysis of Non-Pressure-Compensating and Pressure-Compensating Drip Emitters" by H. Perea, J. Enciso-Medina, V. P. Singh, D. P. Dutta, and B. J. Lesikar (DOI: https://doi.org/10.1061/(ASCE)IR.1943-4774.0000644). J. Irrig. and Drain. Eng., ASCE (2016)
45. H. Anyoji, I.P. Wu, Statistical approach for drip lateral design. Trans. ASAE **30**(1), 187–192 (1987)
46. V.F. Bralts, I.P. Wu, H.M. Gitlin, Manufacturing variation and drip irrigation uniformity. Trans. ASAE **24**(1), 1234–1240 (1981)
47. V.F. Bralts, D.M. Edwards, I.P. Wu, Drip irrigation design and evaluation based on the statistical uniformity concept, in *Advances in Irrigation* (Vol. 4. Academic, New York, 1987). pp. 67–117
48. J.E. Christiansen, *Irrigation by sprinkling*. California Agricultural Experimental Station Bulletin 670 (Univ. of California, Davis, California 1942)
49. S.W. Churchill, Friction factor equation spans all fluid-flow regimes. Chem. Eng. **84**(24), 91–92 (1977)
50. R.H. Cuenca, *Irrigation systems design: an engineering approach* (Prentice Hall, New Jersey, 1989)
51. S.K. Jain, K.K. Singh, R.P. Singh, Microirrigation lateral design using lateral discharge equation. J. Irrig. and Drain. Eng. **128**(2), 125–128 (2002)
52. M.E. Jensen, Design and operation of farm irrigation systems. ASAE Monograph (ASAE St. Joseph, Mich, 1981)
53. M. Yitayew, A.W. Warrick, Trickle lateral hydraulics.II: design and examples. J. Irrig. and Drain. Eng., ASCE **114**(2), 289–300 (1988)
54. G. Yıldırım, Comparative analysis of hydraulic calculation methods in design oh micro-irrigation laterals and linear solution of hydraulic analysis of tapered laterals, PhD. Thesis (Istanbul Technical University, Hydraulics Division, Istanbul, Turkey, 2007)
55. G. Yıldırım, Computer-aided design of micro-irrigation laterals by using LatCAD (Thesis, Istanbul Technical University, Hydraulics Division, Istanbul, Turkey, MSc, 2001)

3 Revised-Energy-Gradient Line (REGL) Method with Evaluating Minor Head Losses Due to Successive Individual Outlets Barbs Connections Using Kinetic Head Reduction Factor Formulation

Introduction

The insertion of emitters along a trickle lateral modifies the flow streamlines, inducing local turbulence that results in additional local pressure losses, sometimes called minor losses, rather than the pipe friction losses. To accurately evaluate total energy losses in the laterals, these minor losses due to the protrusion of emitter connections in the pipe wall that must be added to the friction losses occurring in the pipe [1].

In the past, numerous researches have been done on the hydraulic analysis of trickle irrigation pipeline networks [2–13]. As a matter of fact, a significant amount of these researches [14–21] do not taken into account the effect of local losses in the design procedure, although the importance of minor losses has been recently presented in the experimental analysis [22–28], and the alternative analytical and numerical approaches [1, 24–26, 29–31].

In a recent study, the state-of-the art review of some recent studies for microirrigation lateral design was presented [32, 33]. In their analysis, a detailed comparison test was applied to several design examples with the special limited design conditions of some calculation methods, such as considering effect of minor head losses, emitter discharge exponent, pipe flow regime and, velocity head considerations; then the dimensionless design charts for different ranges of lateral diameter and length and for various uniform line slope cases, are also presented.

This chapter intends to present an improved analytical procedure based on the Revised Energy-Gradient Line (R-EGL) method (Chap. 2) for hydraulic design of trickle laterals which takes into consideration the effect of minor head losses as well as the effect of the emitter outflow nonuniformity, the kinetic head change, the number of emitters, and different uniform line slopes, on the lateral hydraulic computations [12].

G. Yıldırım, *Hydraulic Principles and Design Concepts for Submain Units with Multiple Outlet Pipelines*, Synthesis Lectures on Engineering, Science, and Technology,
https://doi.org/10.1007/978-3-031-32495-6_3

Total Energy Losses Components

Trickle lateral design procedure needs to evaluate accurately both the pipe head losses due to friction and the local pressure losses that are due to the protrusion of emitter barbs into the flow. The local pressure losses, in fact (in relation to the high number of emitters located along the lateral line) can become significant compared to the overall energy loss [22, 23, 34–37].

Head loss computation in a lateral with multiple emitters using analytical methods is normally based on the head loss that occurs in a similar pipeline without emitters along its length multiplied by an appropriate correction factor. The correction factor F_C [38] for the case in which the first emitter is located in a full emitter spacing from the lateral inlet, is of the form

$$F_C = \frac{1}{(m+1)} + \frac{1}{2N} + \frac{\sqrt{m-1}}{6N^2} \tag{3.1a}$$

If the first emitter is half emitter spacing from the lateral inlet, F_C is given by [39]

$$F_C = \frac{2N}{(2N-1)}\left(\frac{1}{m+1} + \frac{\sqrt{m-1}}{6N^2}\right) \tag{3.1b}$$

in which m is defined as velocity (or flow rate) exponent.

It is worthy to note, the F_C by Eq. (3.5) tends to give smaller values with increasing the number of outlets N; and the F_C by Eqs. (3.1a, 3.1b) systematically yields bigger values than those of Eqs. (3.2a, 3.2b, 3.2c). The difference between Eqs. (3.1a, 3.1b) and (3.2a, 3.2b) is negligible as the number of outlets increases. For instance (for m = 1.75), for a relatively large value of N = 150 the difference is 0.6%, whereas for an extreme small value of N = 5, the difference is 12.5%, approximately.

Friction Head Losses

Previous researches showed that though some simple methods of calculating pipe friction head loss (Hazen-Williams, Scobey, i.e.) had been found, the more complex Darcy-Weisbach formula is the most universally accepted for pipe head losses of small diameter polyethylene (*PE*) pipes [26–28, 40].

The general form of the Darcy-Weisbach friction formula can be approximated by the following expression:

$$S_f(l) = \frac{dH_f(l)}{dl} = \frac{f}{D}H_V(l) = \frac{C}{D^n}Q^m(l) \tag{3.2a}$$

where

$$f = c_1[R_I]^{m-2} \tag{3.2b}$$

$$C = \frac{8c_1(\nu/4)^{2-m}}{g\pi^m} \tag{3.2c}$$

in which $S_f(l)$ = friction slope or the slope of the energy grade line at any distance l from the pipeline inlet expressed as the energy loss per unit length of the pipeline-per unit weight of fluid; $H_f(l)$ = friction head loss per unit weight of fluid at any distance l; f = constant friction coefficient for the Darcy-Weisbach formula depends upon the flow regime; C = combined friction coefficient for the Darcy-Weisbach formula; c_1 = constant for the particular friction coefficient formula used; $R_I = (4N\overline{q}/\pi D\nu)$, Reynolds number at the inlet of the lateral; m and n = velocity and diameter exponent, respectively. Generally n is approximately equal to m + 3 [15]; ν = kinematic viscosity of water; and g = acceleration due to gravity.

For laminar flow regime (R < 2,000; R = Reynolds number for pipe flow), m = 1.0, n = 4.0, c_1 = 64 and C = 4.195 × 10^{-6}; for turbulent flow regime in a smooth pipe (3,000 < R ≤ 10^5), m = 1.75, n = 4.75, c_1 = 0.316 (Blasius equation) and C = 7.792 × 10^{-4} and finally for fully turbulent flow regime (10^5 < R < 10^7), m = 1.828, n = 4.828, c_1 = 0.130 and C = 9.589 × 10^{-4}, are taken [41].

The introduction of the Blasius friction coefficient, $f = 0.316\ R^{-0.25}$ (or C = 7.792 × 10^{-4}) into the Darcy-Weisbach formula provides an accurate estimation of the frictional losses produced by turbulent flow inside uniform pipes with low wall roughness and when Reynolds number changes within the range 3,000 < R < 10^5. Most trickle laterals are usually made of smooth *PE* pipes, and their flow regime rarely exceeds these flow conditions.

Integrating Eq. (3.2a) between the limits 0 and l yields [20, 42]

$$H_f(l) = H_{f(L)}\left[1 - (1-i)^{m+1}\right] \tag{3.3a}$$

$$H_{f(L)} = F_C H_{f0} \tag{3.3b}$$

$$H_{f0} = f\frac{L}{D}H_{VI} = C\frac{Q_I^m}{D^n}L \tag{3.3c}$$

where $H_{f(L)}$ = total friction drop at the closed end of the lateral; and H_{f0} = total friction drop in a similar pipeline without emitters conveying the total flow rate along its entire length.

Minor Head Losses Due to Outlet Connections

The presence of an individual emitter in a trickle lateral determines a local pressure loss λ_i, which is frequently expressed as a fraction α_i of the kinetic head [37]:

$$\lambda_i = \alpha_i \frac{V_I^2}{2g} \tag{3.4}$$

Although each individual emitter usually determines a small local head loss given by Eq. (3.4), the sum of all these minor losses in a lateral line can become significant, since emitter spacing can be very small and many emitters are inserted along a lateral line [37].

The total minor head losses along the lateral $H_{s(L)}$, through local singularities produced at emitter insertions of on-line, in-line and integrated emitters, can be evaluated from:

$$H_{s(L)} = F_C N \lambda_i \tag{3.5}$$

Total Head Losses

Total energy losses along the lateral $H_{T(L)}$, includes friction and minor losses can be formulated by the following expression:

$$H_{T(L)} = H_{f(L)} + H_{s(L)} = F_C \left[C \frac{Q_I^m}{D^n} L + \left(\frac{8}{\pi^2 g} \right) N \alpha_i \frac{Q_I^2}{D^4} \right] \tag{3.6}$$

Methodology: Energy Relations with Considering Minor Head and Pressure Losses

As illustrated in Chap. 2 (Fig. 3.1), let define H(l) be the pressure head at a given length l from the lateral inlet, H_I be the pressure head at the lateral inlet, and S_0 be the uniform slope of the lateral line.

Assuming the average pressure head for the entire lateral is located in the middle of the length of the lateral, and applying conservation of energy principle between the pipe segments $l = 0$ and $l = l$, then pressure head distribution along the lateral is expressed by

$$\begin{aligned} H(l) = H_I + H_{VI} \{ [1 - (1-i)^2] + (\psi - 1)[1 - (1-i)^{m+1}] \} \\ - H_{f(L)}[1 - (1-i)^{m+1}] - p i \Delta S \end{aligned} \tag{3.7}$$

The average pressure head along the pipeline, $\overline{H}$ can then be written [2]

$$\overline{H} = H_I - (\overline{H}_f + p\overline{\Delta S}) \tag{3.8}$$

in which $\overline{H}_f$ = average head loss due to friction along the lateral; and $\overline{\Delta S} = \Delta S/2$ = average head loss or gain due to elevation change for the entire lateral.

The average head loss due to friction along the laeral, $\overline{H}_f$ is expressed by [2]

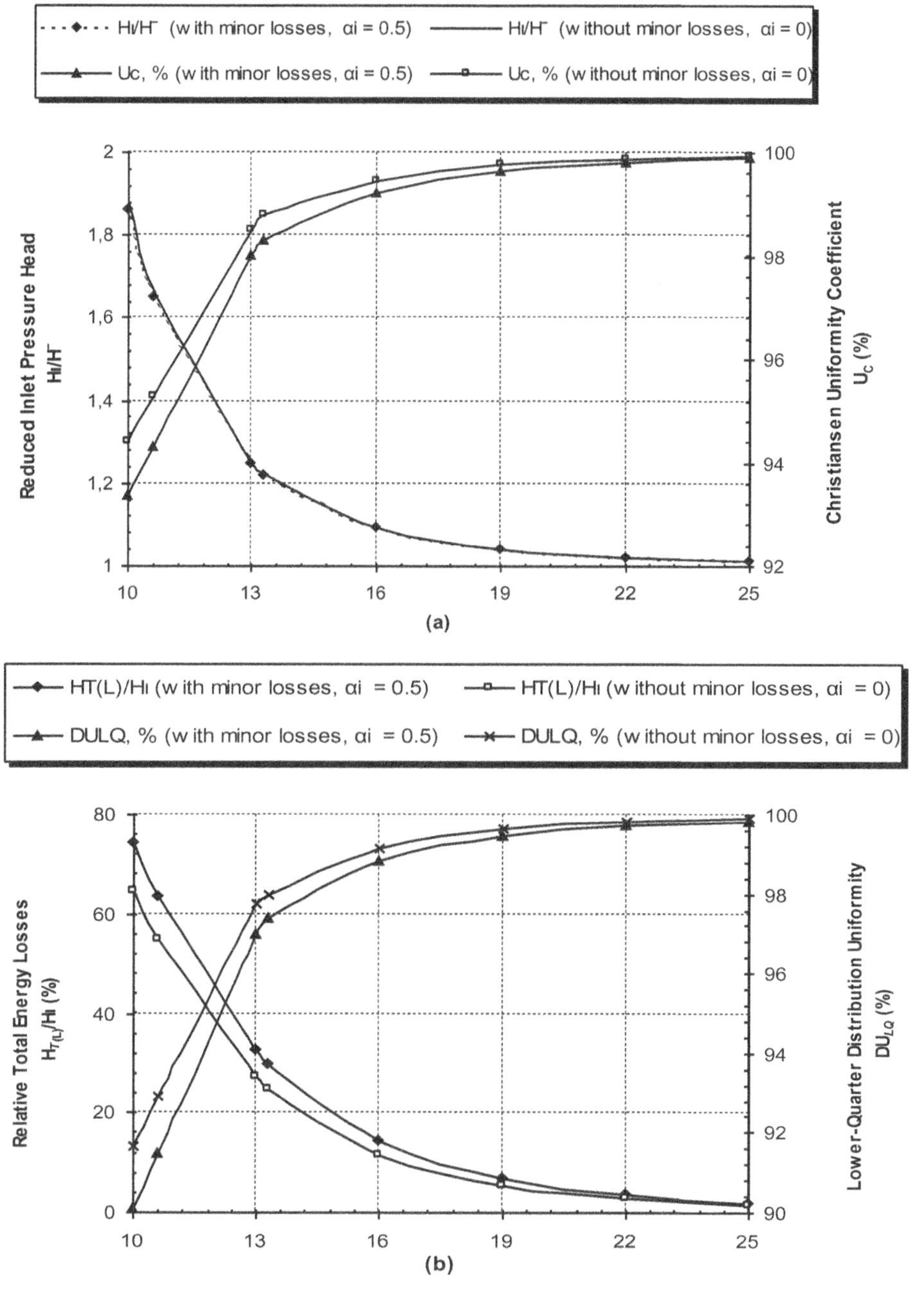

Fig. 3.1 Design chart for the Example 1: for the various lateral diameters, **a** variation of the reduced inlet pressure head and U_C uniformity coefficient and, **b** variation of the relative total energy losses and DU_{LQ} uniformity coefficient

$$\overline{H}_f = \frac{1}{L}\int_{l=0}^{l=L} dS = \frac{1}{L}\int_{l=0}^{l=L} H_f(l)dl = \left(\frac{m+1}{m+2}\right)H_{f(L)} \tag{3.9}$$

Substituting the expression for $\overline{\Delta S}$, and $\overline{H}_f$ by Eq. (3.9) into Eq. (3.8) then, subtracting from Eq. (3.7) finally, the result becomes to Eq. (3.10):

$$\begin{aligned} H(l) = \overline{H} + H_{VI}\{[1-(1-i)^2] + (\psi-1)[1-(1-i)^{m_\beta+1}]\} \\ + H_{f(L)}\left[(1-i)^{m_\beta+1} - \frac{1}{(m_\beta+2)}\right] - p\Delta S\left(i-\frac{1}{2}\right) \end{aligned} \tag{3.10}$$

in which H(l) = pressure head at a given section of length, l from the lateral inlet (m); $\overline{H}$ = average pressure head for the entire of the lateral; $H_{VI} = 8(N\overline{q})^2/\pi^2 g D^4$ = velocity head at the lateral inlet; $m_\beta = \beta \times m$ = improved value of the velocity exponent m; ψ = kinetic head reduction factor that accounts minor losses due to emitter insertions, given by Eq. (3.11); $\Delta S = S_0 L$ = total head loss or gain due to change of elevation at the closed end of the lateral; and S_0 = line slope assuming to be uniform along the lateral.

Noting that in this procedure, the slope of a downhill lateral is negative ($p = -1$), and the slope of an uphill lateral is positive ($p = 1$).

Kinetic Head Reduction Factor (Ψ) Formulation

We illustrate a new term, ψ, into Eqs. (3.7) or (3.10) defined as "kinetic head reduction factor" to take into account the effect of minor losses due to emitter connections to introduce into the kinetic head, which is evaluated by the following expression:

$$\psi = 1 - F_C N \alpha_i \tag{3.11}$$

where α_i = fraction of the kinetic head which is characterized by different kind of emitters; F_C and N are previously defined.

Improved Nonuniform Outflow Exponent (Β)

To alternate the traditional relationship for the linear decreasing flow rate through the lateral line, to describe the variable flow rate along the pipeline Q(l), simple but sufficiently accurate exponential function can be used [21, 40, 43]

$$Q(l) = Q_I(1-i)^\beta \tag{3.12}$$

If Eq. (3.12) is evaluated for the upper half segment of the pipeline (i = 0.5), then arranged with the expression for the flow rate entering over the upper half segment of the lateral, $Q_{0.5} = (kN/2)\overline{H}_{0.5}^x$ (k = coefficient of the emitter discharge) finally, the β can be evaluated from

$$\beta = 1 - 3.322x \log\left(\frac{\overline{H}_{0.5}}{\overline{H}}\right) \tag{3.13}$$

where $\overline{H}_{0.5}$ = average pressure head along the upper half segment of the lateral.

Assuming the $\overline{H}_{0.5}$ is located in the middle of the length of the upper half segment of the lateral, and applying conservation of energy equation between the pipe segments l = L/4 and l = L, then the $\overline{H}_{0.5}$ can be expressed [21, 40]

$$\overline{H}_{0.5} = H_N + H_{f(L)} + (p\overline{\Delta S}_{0.5} - \overline{H}_{f0.5}) \tag{3.14}$$

where H_N = downstream pressure head at the first emitter from the closed end of the lateral; $\overline{H}_{f0.5}$ = average head loss due to friction along the upper half segment of the lateral; and $\overline{\Delta S}_{0.5}$ = average head loss or gain due to elevation change at the end of the upper-quarter segment of the lateral, which is written by this form:

$$\overline{\Delta S}_{0.5} = \frac{3}{4}\Delta S \tag{3.15}$$

The operating inlet pressure head, H_{0I}, can be directly obtained for i = 1, from the pressure head profile given by Eq. (3.10) with taking into account the flow rate exponent, $m_\beta = 1.0 \times m$:

$$H_{0I} = \overline{H} + \frac{(m+1)}{(m+2)}H_{f(L)} + p\frac{\Delta S}{2} \tag{3.16a}$$

Similarly, the downstream end pressure head, H_N can be obtained for i = 1 from the pressure head profile, Eq. (3.10):

$$H_N = \overline{H} + \psi H_{VI} - \left(\frac{1}{m+2}\right)H_{f(L)} - \frac{p\Delta S}{2} \tag{3.16b}$$

Average head loss due to friction between the pipe segments l = 0 and l = L/2 for the upper half segment of the lateral, $\overline{H}_{f0.5}$, is expressed by [34, 35]

$$\overline{H}_{f0.5} = \frac{2}{L}\int_{l=0}^{l=L/2} dS = \frac{2}{L}\int_{l=0}^{l=L/2} H_f(l)dl = \left(\frac{m + 0.5^{m+1}}{m+2}\right)H_{f(L)} \tag{3.17}$$

Substituting Eqs. (3.15), (3.16b) and (3.17) into Eq. (3.14), and dividing by $\overline{H}$ then, substituting into Eq. (3.13) finally, the result transforms to Eq. (3.18):

$$\beta = 1 - 3.322x \log\left[1 + \psi\frac{H_{VI}}{\overline{H}} + \lambda\frac{H_{f(L)}}{\overline{H}} + \frac{1}{4\overline{H}}(p\Delta S)\right]; \lambda = \left(\frac{1 - 0.5^{m+1}}{m+2}\right) \tag{3.18}$$

in which β = nonuniform outflow exponent then, the outflow from each multiple emitter will not be uniform along the pipeline.

Uniformity Criteria for Water Application

Allowable Limit for Extreme Pressure Head Difference

Based on the analytical development presented in Chap. 2, the maximum allowable difference in the outlet operating pressure heads can be written in the general form [21]:

$$\begin{aligned} H_{\max} - H_{\min} = H_{f(L)}\left[(1 - i_{\max})^{m_\beta+1} - (1 - i_{\min})^{m_\beta+1}\right] \\ + H_{VI}\left[(1 - i_{\min})^2 - (1 - i_{\max})^2\right] - p\Delta S(i_{\max} - i_{\min}) \end{aligned} \tag{3.19}$$

In uphill and zero slope conditions, the maximum operating pressure head is always located at the first emitter from the lateral inlet ($i_{\max}$ = μ s/L), whereas the minimum operating pressure head is always found at the first emitter from the downstream closed end of the lateral ($i_{\min}$ = 1.0).

Using the expressions for $i_{\max}$ and $i_{\min}$ in Eq. (3.19) and the result into the expression for δ_H [Eq. (3.32)], finally for uphill slope case, the δ_H can be similarly obtained as follows

$$\delta_H = \frac{\Delta H}{\overline{H}} = \frac{(H_{\max} - H_{\min})}{\overline{H}} = \frac{\left[H_{f(L)}\theta^{m_\beta+1} + H_{VI}\left[(1-\psi)\theta^{m_\beta+1} - \theta^2\right] + p\Delta S\theta\right]}{\overline{H}} \tag{3.20}$$

In downhill lines, the location of the maximum and the minimum operating pressure heads is variable and depends on the relationship between the total energy gain or loss due to uniform slope and the total head losses due to pipe friction. The location of the maximum operating pressure head is found either to the first emitter near the lateral inlet ($i_{\max}$ = μ s/L) or at the first emitter from the downstream closed end point ($i_{\max}$ = 1.0).

In downhill lines, the current design procedure assumes the most usual case in which the maximum operating outlet pressure head occurs at the first emitter near the downstream closed end of the lateral

$$i_{\max} = \frac{l_{\max}}{L} = 1 \tag{3.21a}$$

The location of the minimum operating pressure head is determined where the first derivative of the pressure head distribution given by Eq. (3.10) is equal to zero. For simplification of this derivative procedure, the velocity head term can be omitted in this equation, then the location $i_{\min}$ where the minimum pressure head occurs, can be approximated by

$$i_{\min} = \frac{l_{\min}}{L} = 1 - \left[\frac{(-p\Delta S)}{H_{f(L)}(m_\beta + 1)}\right]^{1/m_\beta} \tag{3.21b}$$

Using Eqs. (3.21a) and (3.21b) for i_{max} and i_{min} into Eq. (3.19) and, the result into the expression for δ_H [Eq. (3.32)]:

$$\delta_H = \frac{\left[H_{VI}\left[(1-i_{min})^2 + (\psi - 1)(1-i_{min})^{m_\beta+1}\right] - H_{f(L)}(1-i_{min})^{m_\beta+1} - p\Delta S(1-i_{min})\right]}{H_{max}} \tag{3.22}$$

where θ = L$_0$/L = length ratio which is the percentage of the distance between the first and last emitters to the full length of the lateral; $L_0 = (N-1)s$ = length of the distance between the first and last emitters except for the inlet spacing, s_I; and $L = (N+\mu-1)s$ = full length of the lateral as defined in the previous section.

Coefficient of Variation of Discharge (CV$_q$), Christiansen's Uniformity (U$_C$) and Lower-Quarter Distribution Uniformity (DU$_{LQ}$)

The Christiansen's uniformity coefficient U$_C$ is evaluated from the following expression including the kinetic head reduction factor, ψ:

$$\begin{aligned} U_C = 1 - 0.798\frac{x}{\overline{H}}\Bigg(& H_{f(L)}^2 \frac{(m_\beta+1)^2}{(m_\beta+2)^2(2m_\beta+3)} \\ & + H_{VI}^2\left\{\frac{2(m_\beta+1)^2}{(m_\beta+2)(2m_\beta+3)}(1-\psi)^2 + \left[\frac{1}{(m_\beta+2)(m_\beta+4)} - \frac{1}{3}\right]4(1-\psi) + \frac{8}{15}\right\} \\ & + 2H_{f(L)}H_{VI}\left[\frac{(m_\beta+1)^2}{(m_\beta+2)^2(2m_\beta+3)}(1-\psi) - \frac{2}{3}\frac{(m_\beta+1)}{(m_\beta+2)(m_\beta+4)}\right] \\ & + H_{VI}p\Delta S\left[\frac{(m_\beta+1)}{(m_\beta+2)(m_\beta+3)}(1-\psi) - \frac{1}{6}\right] \\ & + H_{f(L)}p\Delta S\frac{(m_\beta+1)}{(m_\beta+2)(m_\beta+3)} + \frac{1}{12}(p\Delta S)^2\Bigg)^{1/2} \end{aligned} \tag{3.23}$$

For normal distribution, the coefficient of variation of discharge, δ_q is evaluated from the relationship between U$_C$ and CV$_q$; then the lower-quarter distribution uniformity, D$_{ULQ}$ can be evaluated from the simple transformation between U$_C$ and D$_{ULQ}$.

Computational Example

Determine the percentages of total energy losses for two on-line emitter models (labyrinth and orifice-vortex) considering horizontal trickle lateral having the internal pipe diameter D = 13.37 mm and pipe lengths L = 33, 54, 72 and 90 m, by setting the emitter spacing, s, equal to the commercially available values of 0.3, 0.6, 0.9 and, 1.2 m, respectively. The following experimental relationships with minor loss coefficients for both the emitter models are considered as follows [26]:

Labyrinth on-line emitter model: $q_{(lph)} = 1.32 \times \overline{H}_{(m)}^{0.5}$; $\overline{q} = 4.1$ (lph) ($\alpha_i = 0.34$)

Orifice-vortex on-line emitter model: $q_{(lph)} = 1.65 \times \overline{H}_{(m)}^{0.4}$; $\overline{q} = 4.2$ (lph) ($\alpha_i = 0.67$)

Solution for Labyrinth On-Line Emitter Model ($\alpha_i = 0.34$)***:***

In order to show the applicability of the present technique, the first group for input parameters (s = 0.3 m, N = 110 and L = 33 m) for the type of labyrinth on-line emitter was taken into consideration. The calculation steps are as follows:

1. The required average pressure head, $\overline{H}$ is evaluated from the given emitter outflow-pressure head relationship:

$$\begin{aligned} q_{(lph)} &= 1.32 \times \overline{H}_{(m)}^{0.5}; \overline{q} = 4.1(lph) \\ &= 4.1 \times \left(10^{-3}/3600\right) = 1.139 \times 10^{-6}\,\mathrm{m}^3\,\mathrm{s}^{-1}; x = 0 \end{aligned}$$

$$\overline{H}_{(m)} = \left(\frac{\overline{q}_{(lph)}}{1.32}\right)^{1/0.5} = \left(\frac{4.1}{1.32}\right)^2 = 9.648m$$

2. Total inflow rate for the entire length of the pipeline and inlet flow velocity are approximated by:

$$Q_I = N\overline{q} = 110 \times 1.139 \times 10^{-6} = 1.253 \times 10^{-4}\,\mathrm{m}^3\mathrm{s}^{-1}$$

$$V_I = \frac{4Q_I}{\pi D^2} = \frac{4 \times 1.253 \times 10^{-4}}{\pi \times (13.37 \times 10^{-3})^2} = 0.892\,\mathrm{ms}^{-1}$$

3. The total kinetic head at the pipe inlet is evaluated from:

$$H_{VI} = \frac{V_I^2}{2g} = \frac{0.892^2}{2 \times 9.81} \cong 0.0406\,\mathrm{m}$$

4. The value of Reynolds number at the pipe inlet is computed from the definition of the Reynolds number:

$$\mathrm{R_I} = \frac{4Q_I}{\pi D \upsilon} = \frac{4 \times 1.253 \times 10^{-4}}{\pi \times \left(13.37 \times 10^{-3}\right) \times 1.01 \times 10^{-6}} \cong 11,812.22$$

5. The turbulent flow ($3{,}000 < \mathrm{R_I} = 11{,}812.22 < 10^5$) occurs at the inlet of lateral, therefore the following coefficients/exponents in Eqs. (3.2b) and (3.2c) are used:

$$c_1 = 0.316, m = 1.75, n = 4.75, C = 7.792 \times 10^{-4}$$

The friction coefficient is evaluated from the general expression, Eq. (3.2b):

$$f = c_1[R_I]^{m-2} = 0.316 \times (11,812.22)^{1.75-2} = 0.0303$$

6. For the total number of outlets N = 33, the percentage difference between the F and F_C (or F_{SW}) is about 4.0% (Fig. 3.2). As a preference, the Christiansen's friction correction factor, F_C for $\mu = s_I/s = 1.0$, from Eq. (3.1a):

$$F_C = \frac{1}{(m+1)} + \frac{1}{2N} + \frac{\sqrt{m-1}}{6N^2}$$

$$F_C = \frac{1}{2.75} + \frac{1}{2 \times 33} + \frac{\sqrt{0.75}}{6 \times 33^2} = 0.3682$$

7. Total friction drop in a similar pipeline without outlets, H_{f0} from Eq. (3.3c):

$$H_{f0} = f\frac{L}{D}H_{VI} = C\frac{Q_I^m}{D^n}L$$

$$H_{f0} = 0.0303 \times \frac{33}{(13.37 \times 10^{-3})} \times 0.0406 = 7.792 \times 10^{-4}$$
$$\times \frac{(1.253 \times 10^{-4})^{1.75}}{(13.37 \times 10^{-3})^{4.75}} \times 33 = 3.036 \cong 3.04m$$

8. Total friction drop at the end of the pipeline, $H_{f(L)}$ from Eq. (3.3b):

$$H_{f(L)} = F_C H_{f0} = 0.3682 \times 3.036 \cong 1.118\, m$$

9. Local loss amount for an individual emitter, λ_i, is evaluated from Eq. (3.4):

$$\lambda_i = \alpha_i \frac{V_I^2}{2g} = 0.34 \times \frac{0.892^2}{2 \times 9.81} \cong 0.0138m$$

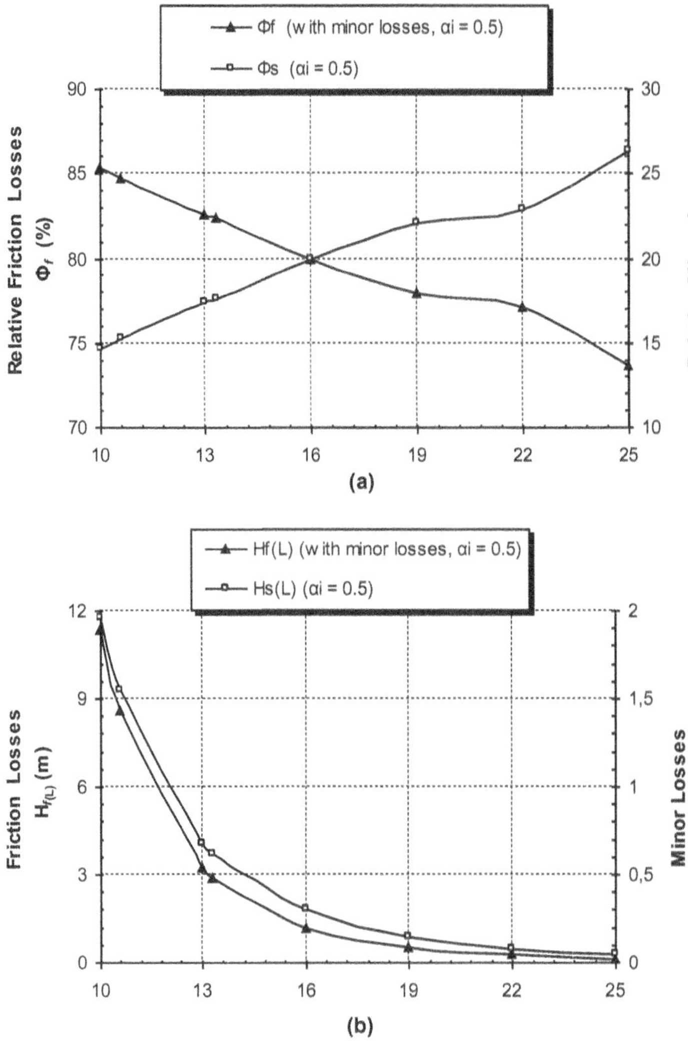

Fig. 3.2 Design chart for the Example 1: for the various lateral diameters, **a** friction and minor losses as percentage of total energy losses and, **b** the amount of friction and minor losses

10. Total amount of local losses due to total number of on-line emitters along the line, $H_{S(L)}$, is evaluated from Eq. (3.5):

$$H_{s(L)} = F_C N \lambda_i = 0.3682 \times 110 \times 0.0138 = 0.559 \cong 0.56m$$

11. Total head losses along the line due to pipe friction and emitter connections, $H_{T(L)}$, is evaluated from:

$$H_{T(L)} = H_{f(L)} + H_{S(L)} = 1.118 + 0.559 = 1.677m$$

12. The kinetic head reduction factor, ψ, is computed from Eq. (3.11):
13. $\psi = 1 - F_C N \alpha_i = 1 - 0.3682 \times 110 \times 0.34 = -12.77$
14. The nonuniform outflow exponent, β, is determined from Eq. (3.18):

$$\beta = 1 - 3.322x \log\left[1 + \psi \frac{H_{VI}}{\overline{H}} + \lambda \frac{H_{f(L)}}{\overline{H}} + \frac{1}{4\overline{H}}(p\Delta S)\right]$$

$$x = 0.5;\ \lambda = \left(\frac{1 - 0.5^{m+1}}{m+2}\right) = \left(\frac{1 - 0.5^{2.75}}{3.75}\right) = 0.227$$

$$\beta = 1 - 3.322 \times 0.5 \times log\left[1 + (-12.77) \times \frac{0.0406}{9.648} + 0.227\right.$$

$$\left. = \times \frac{1.118}{9.648} + \frac{1}{4 \times 9.648} \times (0)\right] 1.02$$

15. The improved value of velocity exponent m_β:

$$m_\beta = \beta \times m = 1.02 \times 1.75 = 1.785$$

16. The operating inlet pressure head, H_{OI}, for nonuniform outflow profile is determined from Eq. (3.16a)

$$H_{OI} = \overline{H} + \frac{(m_\beta + 1)}{(m_\beta + 2)} H_{f(L)} + p\frac{\Delta S}{2} = 9.648 + \frac{(1.785 + 1)}{(1.785 + 2)} \times 1.118 + 0 = 10.471m$$

17. The downstream end pressure head, H_N, from Eq. (3.16b):

$$H_N = \overline{H} + \psi H_{VI} - \left(\frac{1}{m_\beta + 2}\right) H_{f(L)} - \frac{p\Delta S}{2} = 9.648 + (-12.77)$$

$$\times 0.0406 - \frac{1}{(1.785 + 2)} \times 1.118 = 8.834m$$

18. We check the amount of local losses from the conservation of energy principle:

$$H_{T(L)} = (H_{f(L)} + H_{S(L)}) = (H_{0I} - H_N) + H_{VI}$$

$$H_{S(L)} = (H_{0I} - H_N) + (H_{VI} - H_{f(L)})$$

$$H_{S(L)} = (10.471 - 8.834) + (0.0406 - 1.118) = 0.559m$$

Or, directly:

$$H_{0I} - H_N = -\Psi H_{VI} + H_{f(L)}\left(\frac{m_\beta + 1}{m_\beta + 2} + \frac{1}{m_\beta + 2}\right) = -\Psi H_{VI} + H_{f(L)}$$

$$H_{S(L)} = -\Psi H_{VI} + H_{f(L)} + H_{VI} - H_{f(L)} = H_{VI}(1 - \Psi)$$

$$H_{S(L)} = 0.0406 \times [1 - (-12.77)] = 0.0406 \times 13.77 = 0.559m$$

Solution for Orifice-Vortex On-Line Emitter Model ($\alpha_i = 0.67$):
In the following the first group for input parameters (s = 0.3 m, N = 110 and L = 33 m) for the type of orifice-vortex on-line emitter was taken into consideration. The calculation steps are as follows:

1. The required average pressure head, $\overline{H}$ is evaluated from the given emitter outflow-pressure head relationship:

$$q_{(lph)} = 1.65 \times \overline{H}^{0.4}_{(m)}; \quad \overline{q} = 4.2 \quad (lph) = 4.2 \times \left(10^{-3}/3600\right)$$
$$= 1.167 \times 10^{-6}\,\mathrm{m^3s^{-1}};\; x = 0.4$$

$$\overline{H}_{(m)} = \left(\frac{\overline{q}_{(lph)}}{1.65}\right)^{1/0.4} = \left(\frac{4.2}{1.65}\right)^{2.5} = 10.337m$$

2. Total inflow rate for the entire length of the pipeline and inlet flow velocity are approximated by:

$$Q_I = N\overline{q} = 110 \times 1.167 \times 10^{-6} = 1.284 \times 10^{-4}\,\mathrm{m^3s^{-1}}$$

$$V_I = \frac{4Q_I}{\pi D^2} = \frac{4 \times 1.284 \times 10^{-4}}{\pi \times (13.37 \times 10^{-3})^2} = 0.914\,\text{ms}^{-1}$$

3. The total kinetic head at the pipe inlet is evaluated from:

$$H_{VI} = \frac{V_I^2}{2g} = \frac{0.914^2}{2 \times 9.81} \cong 0.0426\,\text{m}$$

4. The value of Reynolds number at the pipe inlet is computed from the definition of the Reynolds number:

$$\text{R}_\text{I} = \frac{4Q_I}{\pi D \nu} = \frac{4 \times 1.284 \times 10^{-4}}{\pi \times (13.37 \times 10^{-3}) \times 1.01 \times 10^{-6}} \cong 12,106.6$$

5. The turbulent flow ($3,000 < \text{R}_\text{I} = 12,106.6 < 10^5$) occurs at the inlet of lateral, therefore the following coefficients/exponents in Eqs. (3.2b) and (3.2c) are used:

$$\text{c}_1 = 0.316, \text{m} = 1.75, \text{n} = 4.75, C = 7.792 \times 10^{-4}$$

The friction coefficient is evaluated from the general expression, Eq. (3.2b):

$$f = c_1[R_I]^{m-2} = 0.316 \times (11,812.22)^{1.75-2} = 0.0301$$

6. The Christiansen's friction correction factor, F_C for $\mu = s_I/s = 1.0$, from Eq. (3.1a):

$$F_C = \frac{1}{(m+1)} + \frac{1}{2N} + \frac{\sqrt{m-1}}{6N^2}$$

$$F_C = \frac{1}{2.75} + \frac{1}{2 \times 33} + \frac{\sqrt{0.75}}{6 \times 33^2} = 0.3682$$

7. Total friction drop in a similar pipeline without outlets, H_{f0} from Eq. (3.3c):

$$H_{f0} = f\frac{L}{D}H_{VI} = C\frac{Q_I^m}{D^n}L$$

$$H_{f0} = 0.0301 \times \frac{33}{(13.37 \times 10^{-3})} \times 0.0426 = 7.792 \times 10^{-4}$$
$$\times \frac{(1.284 \times 10^{-4})^{1.75}}{(13.37 \times 10^{-3})^{4.75}} \times 33 = 3.036 \cong 3.165m$$

8. Total friction drop at the end of the pipeline, $H_{f(L)}$ from Eq. (3.3b):

$$H_{f(L)} = F_C H_{f0} = 0.3682 \times 3.165 \cong 1.165m$$

9. Local loss amount for an individual emitter, λ_i, is evaluated from Eq. (3.4):

$$\lambda_i = \alpha_i \frac{V_I^2}{2g} = 0.67 \times \frac{0.914^2}{2 \times 9.81} \cong 0.0285\,\text{m}$$

10. Total amount of local losses due to total number of on-line emitters along the line, $H_{S(L)}$, is evaluated from Eq. (3.5):

$$H_{s(L)} = F_C N \lambda_i = 0.3682 \times 110 \times 0.0285 = 0.559 \cong 1.155\,\text{m}$$

11. Total head losses along the line due to pipe friction and emitter connections, $H_{T(L)}$, is evaluated from:

$$H_{T(L)} = H_{f(L)} + H_{S(L)} = 1.165 + 1.155 = 2.320\,\text{m}$$

12. The kinetic head reduction factor, ψ, is computed from Eq. (3.11):

$$\psi = 1 - F_C N \alpha_i = 1 - 0.3682 \times 110 \times 0.67 = -26.14$$

13. The nonuniform outflow exponent, β, is determined from Eq. (3.18):

$$\beta = 1 - 3.322x \log\left[1 + \psi\frac{H_{VI}}{\overline{H}} + \lambda\frac{H_{f(L)}}{\overline{H}} + \frac{1}{4\overline{H}}(p\Delta S)\right]$$

$$x = 0.4;\ \lambda = \left(\frac{1 - 0.5^{m+1}}{m+2}\right) = \left(\frac{1 - 0.5^{2.75}}{3.75}\right) = 0.227$$

$$\beta = 1 - 3.322 \times 0.4 \times \log[1 + (-26.14) \times \frac{0.0426}{10.337} + 0.227 \times \frac{1.165}{10.337} + \frac{1}{4 \times 10.337} \times (0)] = 1.0495$$

14. The improved value of velocity exponent m_β:

$$m_\beta = \beta \times m = 1.0495 \times 1.75 = 1.837$$

15. The operating inlet pressure head, H_{0I}, for nonuniform outflow profile is determined from Eq. (3.16a)

$$H_{0I} = \overline{H} + \frac{(m_\beta + 1)}{(m_\beta + 2)}H_{f(L)} + p\frac{\Delta S}{2} = 10.337 + \frac{(1.837 + 1)}{(1.837 + 2)} \times 1.165 + 0 = 11.198m$$

16. The downstream end pressure head, H_N, from Eq. (3.16b):

$$H_N = \overline{H} + \psi H_{VI} - \left(\frac{1}{m_\beta + 2}\right)H_{f(L)} - \frac{p\Delta S}{2} = 10.337 + (-26.14) \times 0.0426 - \frac{1}{(1.837 + 2)} \times 1.165 = 8.92m$$

17. We check the amount of local losses from the conservation of energy principle:

$$H_{T(L)} = (H_{f(L)} + H_{S(L)}) = (H_{0I} - H_N) + H_{VI}$$

$$H_{S(L)} = (H_{0I} - H_N) + (H_{VI} - H_{f(L)})$$

$$H_{S(L)} = (11.198 - 8.92) + (0.0426 - 1.165) = 1.156m$$

Or, directly:

$$H_{0I} - H_N = -\Psi H_{VI} + H_{f(L)}\left(\frac{m_\beta + 1}{m_\beta + 2} + \frac{1}{m_\beta + 2}\right) = -\Psi H_{VI} + H_{f(L)}$$

$$H_{S(L)} = -\Psi H_{VI} + H_{f(L)} + H_{VI} - H_{f(L)} = H_{VI}(1 - \Psi)$$

$$H_{S(L)} = 0.0426 \times [1 - (-26.14)] = 0.0426 \times 27.14 = 1.156m$$

The semi-theoretical solution presented-above was extended for both the emitter models covering given input parameters, and the complete results were shown in Table 3.1. This table illustrates for both the emitter models, for different emitter spacing, s, the number of emitters, N, and total length for the entire of the lateral pipeline L; the values of the kinetic head reduction factor which accounts the effect of minor losses, ψ, the amount of minor losses, $H_{s(L)}$, the amount of friction losses in the pipe, $H_{f(L)}$, together with minor losses, Φ_s, expressed as a percentage of total energy losses, respectively.

Table 3.1 indicates that for both the emitter models examined, the length of the lateral is increasing with increasing emitter spacing, whereas the number of emitters is decreasing. As a consequence of reducing the number of emitters and increasing the length of the lateral, the amount of minor losses is decreasing with increasing emitter spacing, whereas the pipe friction losses are increasing. For a certain trickle lateral design, therefore, the amount of minor losses, Φ_s, expressed as a percentage of the total energy losses, is decreasing with increasing emitter spacing, with values that can reach 33.3% and 50.0%, respectively, for the labyrinth and orifice-vortex emitter models, with the highest Φ_s value obtained for the emitter model characterized by the highest α_i.

Table 3.1 Minor losses as percentage of total energy losses obtained for two on-line emitter models and for different emitter spacing

Design parameters			Labyrinth on-line emitter model ($\alpha_i = 0.34$)				Orifice-vortex on-line emitter model ($\alpha_i = 0.67$)			
s (m)	N (–)	L[a] (m)	ψ[b] (–)	$H_{s(L)}$ (m)	$H_{f(L)}$ (m)	Φ_s[c] (%)	ψ[b] (–)	$H_{s(L)}$ (m)	$H_{f(L)}$ (m)	Φ_s[c] (%)
0.3	110	33.0	−12.77	0.559	1.118	33.3	−26.14	1.156	1.165	50.0
0.6	90	54.0	−10.29	0.307	1.277	19.4	−21.26	0.635	1.332	32.3
0.9	80	72.0	−9.061	0.216	1.386	13.5	−18.83	0.447	1.446	23.6
1.2	75	90.0	−8.442	0.178	1.548	10.3	−17.61	0.368	1.614	18.6

[a]Full length of the pipeline including inlet spacing from the first emitter to pipe inlet
[b]Kinetic head reduction factor that accounts the effect of minor losses
[c]Minor losses as percentage of total energy losses

Design Examples

In the proposed analytical procedure, the following data are assigned in advance for a certain design: the internal diameter of the lateral D, the uniform slope along the lateral line S_O, total number of operating emitters N, common emitter spacing s, inlet spacing ratio μ, minor loss coefficient α_i, which depends on the emitter model examined, coefficient and exponent of emitter discharge k and x, constants in the Darcy-Weisbach formula (c_1, m and n), and finally the average emitter discharge $\overline{q}$ and corresponding pressure head, $\overline{H}$. In the following section the proposed algorithm presented here, is applied on two numerical examples for designing the diameter and length of lateral, for both the design cases of without and including minor losses.

Example 1: Determination of Lateral Diameter Using Dimensionless Design Curves It is required to design the internal diameter, D, of a horizontal trickle lateral, as well as the pressure head at the inlet, H_I, and the amount of friction and minor losses $H_{f(L)}$ and $H_{s(L)}$ as percentage of total losses, $\Phi_f = H_{f(L)}/H_{T(L)}$ and $\Phi_s = H_{s(L)}/H_{T(L)}$, considering both the design cases (**a**) **without minor losses** ($\alpha_i = 0$) for $U_C = 95.3\%$ and, (**b**) **including minor losses** ($\alpha_i = 0.5$) for $U_C = 98.3\%$, using the following data very similar to the one presented [11]: the number of emitters, N = 101; the spacing between successive emitters, s = 1.0 m with the inlet spacing ratio, $\mu = s_I/s = 1.0$; the emitter discharge exponent, x = 0.2; the average emitter discharge $\overline{q} = 4.0$ lph ($=1.111 \times 10^{-6}$ m^3s^{-1}) and corresponding pressure head, $\overline{H} = 9.6$ m, respectively.

On the basis of the algorithm discussed, the values of the reduced inlet pressure head, $H_I/\overline{H}$ with the Christiansen uniformity coefficient, U_C, and; the relative total energy losses, $H_{T(L)}/H_I$ with the lower-quarter distribution uniformity coefficient, DU_{LQ} are evaluated for different pipe diameters ranging between 10 and 25 mm and, the results are indicated in Fig. 3.1a, b, respectively; and for the design case (**b**) the percentage values of the friction and minor losses, Φ_f and Φ_s and, their real values, $H_{f(L)}$ and $H_{s(L)}$ versus the different diameter values are also indicated in Fig. 3.2a, b, respectively.

In the following, the design solutions for both the design cases (**a**) and (**b**) using the design charts (Figs. 3.1 and 3.2), will be presented.

Design Case (a): Without Minor Losses ($\alpha_i = 0$)

From Fig. 3.1a, for $U_C = 95.3\%$ we have the design values: D = 10.6 mm and, $H_I/\overline{H} = 1.67$, therefore $H_I = 1.67 \times 9.6 = 16.03$ m. From Fig. 3.1b, for D = 10.6 mm the $DU_{LQ} = 92.9\%$ and $H_{T(L)}/H_I = 54.7\%$, therefore $H_{T(L)} = H_{f(L)} = 0.547 \times 16.03 = 8.81$ m, are found.

Design Case (B): Including Minor Losses ($\alpha_i = 0.5$)

From Fig. 3.1a, for $U_C = 98.3\%$ we have D = 13.3 mm and, $H_I/\overline{H} = 1.225$, therefore $H_I = 1.225 \times 9.6 = 11.76$ m. From Fig. 3.1b, for D = 13.3 mm the $DU_{LQ} = 97.4\%$ and $H_{T(L)}/H_I = 29.7\%$, therefore $H_{T(L)} = H_{f(L)} + H_{s(L)} = 0.297 \times 11.76 = 3.52$ m, are found.

From Fig. 3.2a and for D = 13.3 mm, $\Phi_f = H_{f(L)}/H_{T(L)} = 82.4\%$ and, $\Phi_s = H_{s(L)}/H_{T(L)} = 17.6\%$, therefore $H_{f(L)} = 0.824 \times 3.52 = 2.90$ m and, $H_{s(L)} = 0.176 \times 3.52 = 0.62$ m, can be obtained from Fig. 3.2b.

Example 2: Determination of Lateral Length Using Dimensionless Design Curves It is required to design the length, L, of a horizontal trickle lateral, as well as the pressure head at the inlet, H_I, and the amount of friction and minor losses $H_{f(L)}$ and $H_{s(L)}$ as percentage of total energy losses, $\Phi_f = H_{f(L)}/H_{T(L)}$ and $\Phi_s = H_{s(L)}/H_{T(L)}$, considering the both design cases (a) without minor losses ($\alpha_i = 0$) for $U_C = 95.2\%$ and, (b) including minor losses ($\alpha_i = 0.5$) for $U_C = 95.1\%$ using the following data very similar to the one presented by Hathoot et al. [11]: internal diameter of the lateral pipe, D = 13 mm; the spacing between successive emitters, s = 1.0 m with the inlet spacing ratio, $\mu = s_I/s = 1.0$; the emitter discharge exponent, x = 0.53; the average emitter discharge $\overline{q} = 8.0$ lph ($= 2.222 \times 10^{-6}$ m^3s^{-1}) and corresponding pressure head, $\overline{H} = 9.6$ m, respectively.

For the lateral length ranging between 50 and 110 m, $H_I/\overline{H}$ with the corresponding values of U_C and, $H_{T(L)}/H_I$ with the corresponding values of DU_{LQ} are evaluated and, the results are shown in Fig. 3.3a, b, respectively; and for the design case (**b**), the percentage values of total energy losses, Φ_f and Φ_s and, their real values, $H_{f(L)}$ and $H_{s(L)}$ versus the various length values are also shown in Fig. 3.4a, b, respectively.

In the following, the design solutions for both the design cases (**a**) and (**b**) using the design charts (Figs. 3.3 and 3.4), will be presented.

Design Case (A): Without Minor Losses ($\alpha_i = 0$)

As shown in Fig. 3.3a, for $U_C = 95.2\%$ we have the design values: L = 68.0 m and, $H_I/\overline{H} = 1.29$, therefore $H_I = 1.29 \times 9.6 = 12.38$ m. From Fig. 3.3b, for L = 68.0 m the $DU_{LQ} = 92.8\%$ and $H_{T(L)}/H_I = 31.1\%$, therefore $H_{T(L)} = H_{f(L)} = 0.311 \times 12.38 = 3.85$ m, are found.

Design Case (B): Including Minor Losses ($\alpha_i = 0.5$)

As shown in Fig. 3.4a, for $U_C = 95.1\%$ the design length L = 63.0 m and, $H_I/\overline{H} = 1.23$, therefore $H_I = 1.23 \times 9.6 = 11.79$ m. From Fig. 3.4b, for L = 63.0 m the $DU_{LQ} = 92.7\%$ and $H_{T(L)}/H_I = 31.1\%$, therefore $H_{T(L)} = H_{f(L)} + H_{s(L)} = 0.311 \times 11.79 = 3.67$ m, are found.

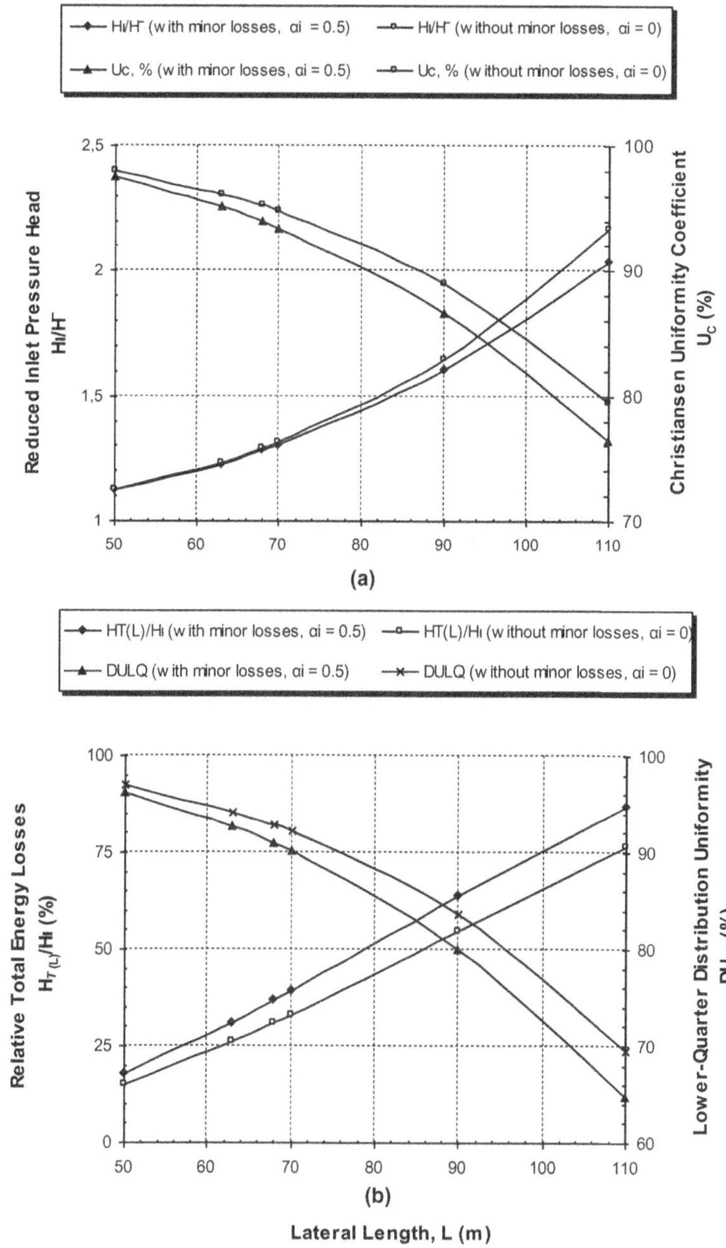

Fig. 3.3 Design chart for the Example 2: for the various lateral lengths, **a** variation of the reduced inlet pressure head and U_C uniformity coefficient and, **b** variation of the relative total energy losses and DU_{LQ} uniformity coefficient

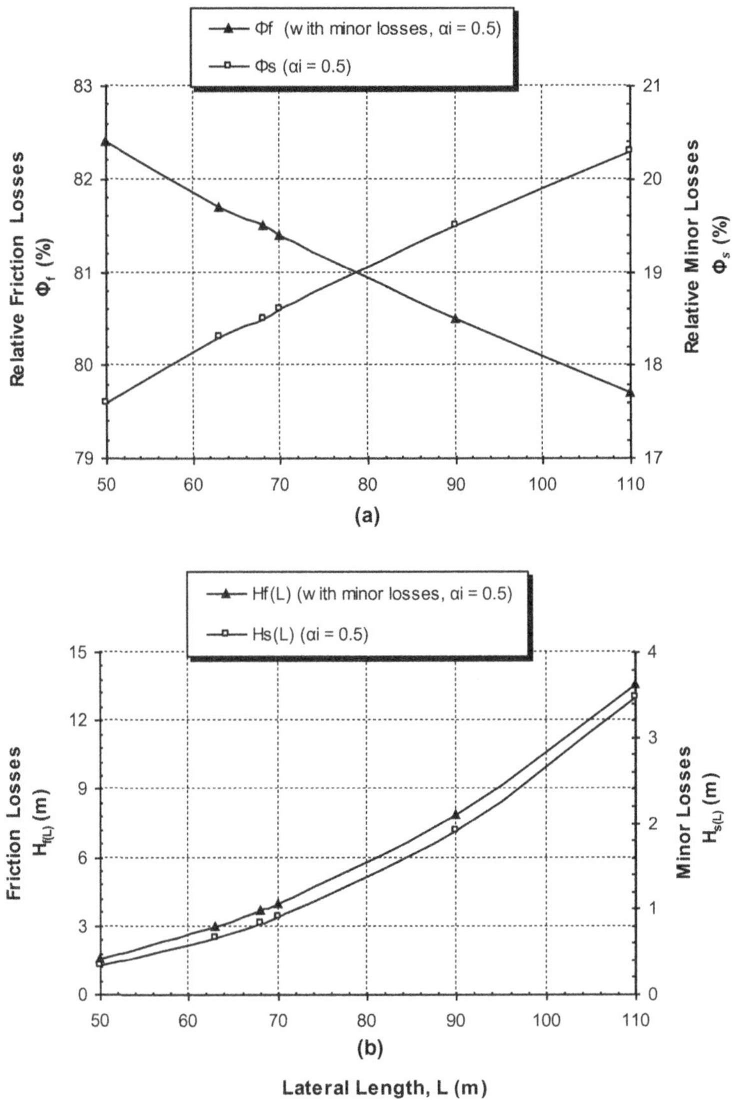

Fig. 3.4 Design chart for the Example 2: for the various lateral lengths, **a** friction and minor losses as percentage of total energy losses and, **b** the amount of friction and minor losses

From Fig. 3.4a and for L = 63.0 m, $\Phi_f = H_{f(L)}/H_{T(L)} = 81.7\%$ and, $\Phi_s = H_{s(L)}/H_{T(L)} = 18.3\%$, therefore $H_{f(L)} = 0.817 \times 3.67 = 2.998$ m and, $H_{s(L)} = 0.183 \times 3.67 = 0.672$ m, can be obtained from Fig. 3.4b.

Comparison Test with Revised Step-By-Step (RSBS) Procedure

In order to illustrate the practical value of the presented analytical solution, the results obtained for the design examples (Examples 1 and 2) are compared with those obtained using the R-SBS design procedure [11] which considers all emitter local losses. Noting that, the authors [11] have used the same data of Examples 1 and 2 for evaluating effect of minor losses, with taking into account various barb sizes (small, medium and large) of an on-line emitter model, whereas the proposed design solution considers an average value of minor loss coefficient, $\alpha_i = 0.5$, to characterize the effect of minor losses for two on-line emitter models (orifice-vortex and labyrinth). In order to do a comparative analysis, the complete results for the design examples obtained from both the procedures are tabulated for both the design cases (**a**) and (**b**), and shown in Tables 3.2 and 3.3, respectively.

Comparative Analysis for Design Example 1

As shown in Table 3.2, for determining lateral diameter, for both the design case (**a**): **without minor losses** and, the design case (**b**): **including minor losses**; the proposed method yields 10.6 mm and 13.3 mm, respectively, whereas the R-SBS method yields 10.0 mm and 13.0 mm, respectively, for three emitter barb sizes, as well. It is evident from this table, considering minor losses (the design case **b**) has the effect of increasing the lateral diameter 25.5% approximately, for the proposed method, whereas 30% for the R-SBS method. However, the uniformity coefficient, U_C has a little change due to the effect of minor losses, according to both the methods.

As shown in Fig. 3.1a and b, as the pipe diameter D increases, the inlet pressure head, H_I and the total energy losses, $H_{T(L)}$ are decreasing, whereas the uniformity coefficients, U_C and DU_{LQ}, are increasing. As a consequence of increasing the diameter values, both the percentage values of the friction head loss, Φ_f (Fig. 3.2a) and its real values, $H_{f(L)}$ (Fig. 3.2b) are decreasing, whereas the percentage values of the minor head loss, Φ_s (Fig. 3.2a) are increasing and its real values, $H_{s(L)}$ (Fig. 3.2b) are decreasing.

Table 3.2 Values of lateral diameter and U_C uniformity coefficient for both the design cases of without and including minor losses according to the proposed analytical method and the revised Step-By-Step (RSBS) method

Design parameters	Proposed analytical method			Revised step-by-step (RSBS) method						
	(a) Without minor losses ($\alpha_i = 0$)	(b) Including minor losses ($\alpha_i = 0.5$)	(a) ~ (b) Increase (%)	Without minor losses	Including minor losses					
					Small barb	Medium barb	Large barb	Increase (%)		
								Small barb	Medium barb	Large barb
D (mm)	10.6	13.3	25.5	10.0	13.0	13.0	13.0	30.0	30.0	30.0
U_C (%)	95.3	98.3	3.15	95.3	98.3	98.3	98.1	3.15	3.15	2.94

Table 3.3 Values of lateral length and U_C uniformity coefficient for both the design cases of without and including minor losses according to the proposed analytical method and the revised Step-By-Step (RSBS) method

Design parameters	Proposed analytical method			Revised step-by-step (RSBS) method						
	(a) Without minor losses ($\alpha_i = 0$)	(b) Including minor losses ($\alpha_i = 0.5$)	(a) ~ (b) Decrease (%)	Without minor losses	Including minor losses					
					Small barb	Medium barb	Large barb	Decrease (%)		
								Small barb	Medium barb	Large barb
L (m)	68.0	63.0	7.4	71.0	68.0	67.0	64.0	4.2	5.6	9.9
U_C (%)	95.2	95.1	0.11	95.2	95.1	95.1	95.1	0.11	0.11	0.11

In order to do a detailed analysis related to the Figs. 3.1 and 3.2, variation of the flow parameters with respect to the lateral diameters and their percentage differences for both the design cases (**a**) and (**b**) are also presented in Table 3.4. It is evident from this table, the minor losses decrease with increasing the pipe diameters. Note that, this relationship was earlier reported in the recent experimental study in which eight different types of on-line emitters were examined [1]. As a remarkable result of this relationship, the percentage increases in total energy losses between the design cases (**a**) and (**b**), increases with increasing the diameter values. In other word, as the pipe diameter increases, for the design case (**b**), the amount of total energy losses decreases due to the decreasing in both friction and minor losses; whereas, for the design case (**a**), the amount of total energy losses have a little decreasing which only depends on the decreasing in pipe friction losses.

Comparative Analysis for Design Example 2

From Table 3.3, for determining the lateral length, for both the design cases (**a**) **without minor losses** and, (**b**) **including minor losses**; the proposed method yields 68.0 m and 63.0 m, respectively, whereas the R-SBS method yields 71 m for the design case (**a**) and, for the design case (**b**); 68.0 m for small barb size, 67.0 m for medium barb size and, 64.0 m for large barb size, respectively. It is obvious from this table, for the proposed method, the effect of minor losses decreases the lateral length 7.4%, approximately, whereas for the R-SBS method, the maximum effect is found with the large size emitter barb where the lateral length is 10% shorter than those of the design case where minor losses are neglected (the design case a). As a consequence, the general results obtained from the proposed method are justified with those obtained from the R-SBS method, for both the design cases (**a**) and (**b**).

As shown in Fig. 3.3a, b, as increasing the lateral length, the inlet pressure head and the total head losses are increasing whereas the uniformity coefficients are decreasing. As a consequence of increasing the length values, both the percentage values of the minor head loss (Fig. 3.4a) and its real values (Fig. 3.4b) are increasing. It should be considered that, as increasing the lateral length, the percentage values of the friction loss (Fig. 3.4a) are decreasing whereas its real values (Fig. 3.4b) are increasing.

Furthermore, variation of the flow variables related to the Figs. 3.3 and 3.4, and their percentage differences for the both design cases (**a**) and (**b**), with respect to the lateral lengths, are also shown in Table 3.5. From this table it is obvious that, the minor losses increase with increasing the pipe length in relation to the high number of emitters located along the lateral line. From this point of view, the percentage increases in total energy losses between the design cases (**a**) and (**b**), decreases with increasing the length values. In other word, as the pipe length increases, for the design case (**b**), the amount of total

Table 3.4 Flow parameters and their percentage differences for both the design cases of without and including minor losses for the various lateral diameters

Lateral Internal Diameter D (mm)	Flow parameters for the design case[a]: without minor losses ($\alpha_i = 0$)						Flow parameters for the design case (b): including minor losses ($\alpha_i = 0.5$)										Differences between the design cases (a) and (b)				
	H_I (m)	H_{VI} (m)	H_N (m)	$H_{f(L)} = H_{T(L)}$ (m)	U_C (%)	DU_{LQ} (%)	H_I (m)	H_{VI} (m)	H_N (m)	$H_{f(L)}$ (m)	$H_{s(L)}$ (m)	$H_{T(L)}$ (m)	Φ_f (%)	Φ_s (%)	U_C (%)	DU_{LQ} (%)	Increase in (%)	Decrease in (%)			
																	$H_{T(L)}$	H_I	H_N	U_C	DU_{LQ}
10[a]	18.14	0.104	6.50	11.74	94.4	91.7	17.94	0.104	4.71	11.37	1.96	13.33	85.3	14.7	93.4	90.1	13.54	1.10	27.54	1.06	1.74
10.6[b]	16.04	0.082	7.30	8.81	95.3 (a), b)	92.9	15.91	0.082	5.86	8.59	1.55	10.13	84.7	15.3	94.3	91.5	14.98	0.81	19.73	1.05	1.51
13[a]	12.02	0.036	8.79	3.27	98.5	97.7	12.00	0.036	8.13	3.23	0.68	3.91	82.6	17.4	98.0	97.0	19.57	0.17	7.51	0.51	0.72
13.3[b]	11.78	0.033	8.88	2.93	98.8	98.0	11.76	0.033	8.27	2.90	0.62	3.52	82.4	17.6	98.3[(a),(b)]	97.4	20.14	0.17	6.87	0.51	0.61
16	10.52	0.016	9.32	1.21	99.4	99.2	10.52	0.016	9.03	1.20	0.30	1.50	80.0	20.0	99.2	98.9	23.97	0.0	3.11	0.20	0.30
19	10.02	0.008	9.50	0.53	99.8	99.6	10.02	0.008	9.35	0.53	0.15	0.68	77.9	22.1	99.6	99.4	28.30	0.0	1.58	0.20	0.20
22	9.83	0.004	9.57	0.27	99.9	99.8	9.83	0.004	9.48	0.27	0.08	0.35	77.1	22.9	99.8	99.7	29.63	0.0	0.94	0.10	0.10
25	9.74	0.003	9.60	0.14	99.9	99.9	9.74	0.003	9.55	0.14	0.05	0.19	73.7	26.3	99.9	99.8	35.71	0.0	0.52	0.0	0.10

[a]Related values obtained from the revised step-by-step (R-SBS) method
[b]Related values calculated from the proposed analytical procedure

Table 3.5 Flow parameters and their percentage differences for both the design cases of without and including minor losses for the various lateral lengths

Lateral length L (m)	Flow parameters for the design case[a]: without minor losses ($\alpha_i = 0$)						Flow parameters for the design case[b]: including minor losses ($\alpha_i = 0.5$)										Differences between the design cases [a, b]				
	H_I (m)	H_{VI} (m)	H_N (m)	$H_{f(L)} = H_{T(L)}$ (m)	U_C (%)	DU_{LQ} (%)	H_I (m)	H_{VI} (m)	H_N (m)	$H_{f(L)}$ (m)	$H_{s(L)}$ (m)	$H_{T(L)}$ (m)	Φ_f (%)	Φ_s (%)	U_C (%)	DU_{LQ} (%)	Increase in (%)	Decrease in (%)			
																	$H_{T(L)}$	H_I	H_N	U_C	DU_{LQ}
50	10.78	0.036	9.20	1.62	97.9	96.9	10.77	0.036	8.87	1.59	0.34	1.93	82.4	17.6	97.5	96.2	19.14	0.10	3.59	0.41	0.72
63[b]	11.85	0.057	8.81	3.10	96.1	94.2	11.79	0.057	8.19	3.0	0.67	3.67	81.7	18.3	95.1[(a),(b)]	92.7	18.39	0.51	7.04	1.04	1.59
64[a]	11.95	0.059	8.77	3.24	95.9	93.9	11.89	0.059	8.12	3.13	0.70	3.83	81.7	18.3	94.9	92.4	18.21	0.51	7.41	1.04	1.59
68[b]	12.39	0.066	8.61	3.85	95.2[(a),(b)]	92.8	12.31	0.066	7.84	3.69	0.84	4.53	81.5	18.5	93.9	90.9	17.66	0.65	8.94	1.37	2.05
70	12.62	0.070	8.52	4.18	94.8	92.2	12.53	0.070	7.69	4.0	0.91	4.91	81.4	18.6	93.4	90.2	17.46	0.71	9.74	1.48	2.17
71[a]	12.75	0.072	8.47	4.35	94.5	91.8	12.64	0.072	7.61	4.15	0.95	5.10	81.4	18.6	93.1	89.8	17.24	0.86	10.15	1.48	2.17
90	15.80	0.116	7.24	8.68	89.0	83.6	15.41	0.116	5.68	7.92	1.92	9.84	80.5	19.5	86.6	79.9	13.36	2.47	21.55	2.69	4.43
110	20.82	0.173	5.01	15.98	79.5	69.5	19.58	0.173	2.72	13.57	3.46	17.03	79.7	20.3	76.4	64.8	6.57	5.96	45.71	3.90	6.76

[a]Related values obtained from the revised step-by-step (R-SBS) method
[b]Related values calculated from the proposed analytical procedure

energy losses increases due to the increasing in both friction and minor losses; whereas, for the design case (**a**), the amount of total energy losses have a little increases which only depends on the increasing in pipe friction losses.

Concluding Remarks

In this chapter, an analytical procedure for hydraulic analysis and solving design problem of a single-diameter trickle lateral, is presented. The proposed equations, taking into consideration the influence of local energy loss (i.e., minor loss), can be applied for various types of emitter, different flow regimes and different uniform line slope ranges. For any desired uniformity level, the procedure provides an opportunity to evaluate the influence of local energy loss, on the pipe geometric characteristics (pipe size and length), and on the corresponding hydraulic variables (operating inlet pressure head, downstream end pressure head and total energy loss). It may be concluded that, the performance of the presented technique is found to be sufficiently accurate in all cases examined even for uphill slope condition in comparison with a revised step-by-step (R-SBS) method.

The procedure can be applied easily for practical purposes without an excessive calculation effort, and preferred instead for the numerical techniques which need a very long execution time and complex computer operation.

Based on the presented research, the following conclusions are underlined:

1. The minor losses due to the emitter insertions increases with decreasing the lateral diameter and increasing the lateral length in relation to the high number of emitters located along the lateral line.
2. Reducing the number of emitters and increasing the length of the lateral, the amount of minor losses decreases with increasing emitter spacing, whereas the pipe friction losses increases. For a certain trickle lateral design, therefore, the amount of minor losses, Φ_s, expressed as a percentage of the total energy losses decreases with increasing emitter spacing, with values that can reach 33.3 and 50.0%, for the labyrinth and orifice-vortex emitter models, respectively, with the highest Φ_s value obtained for the emitter model characterized by the highest value of the minor loss coefficient, α_i.
3. Examination of the results of two practical design examples shows that neglecting the effect of minor losses may lead to design diameters in error by 25% and lengths in error by 7%, approximately.

Acknowledgements The senior author wants to express his acknowledges to Prof. Vijay. P. Singh at Texas A&M University, Emeritus Prof. Necati Ağıralioğlu from International Antalya Science University, and Distinguished Emeritus Professor Zekai Şen from Istanbul Medipol University, for their highly encouragement on preparing this book, many scientific activities, projects and so on. The Scientific and Technological Research Council of Turkey (TUBITAK) is acknowledged for supporting by the fellowships and grant program during the studies in the Texas A&M University/BAEN. We also thank to publishers, potential readers and reviewers for providing constructive reviews and criticism on the content of the book, in advance.

Conflict of Interest The author/editor has no conflicts of interest, financial or otherwise, to declare.

Appendices

Appendix 1: Determination of Energy-Gradient Line (EGL)

Let H(l) be the pressure head at a given length l from the lateral inlet, H_I be the pressure head at the lateral inlet, and S_0 be the uniform slope of the lateral line (Fig. 3.1).

Assuming the average pressure head for the entire lateral is located in the middle of the length of the lateral, and applying conservation of energy principle between the pipe segments $l = 0$ and $l = l$, then pressure head distribution along the lateral is expressed by

$$\begin{aligned} H(l) = H_I + H_{VI}\left\{\left[1-(1-i)^2\right]+(\psi-1)\left[1-(1-i)^{m+1}\right]\right\} \\ - H_{f(L)}\left[1-(1-i)^{m+1}\right] - pi\Delta S \end{aligned} \tag{3.24}$$

The average pressure head along the pipeline, $\overline{H}$ can then be written [21, 40, 42]

$$\overline{H} = H_I - (\overline{H}_f + p\overline{\Delta S}) \tag{3.25}$$

in which $\overline{H}_f$ = average head loss due to friction along the lateral; and $\overline{\Delta S} = \Delta S/2$ = average head loss or gain due to elevation change for the entire lateral [44–47].

The average head loss due to friction along the laeral, $\overline{H}_f$ is expressed by [42, 47]:

$$\overline{H}_f = \frac{1}{L}\int_{l=0}^{l=L} dS = \frac{1}{L}\int_{l=0}^{l=L} H_f(l)dl = \left(\frac{m+1}{m+2}\right)H_{f(L)} \tag{3.26}$$

Substituting the expression for $\overline{\Delta S}$, and $\overline{H}_f$ by Eq. (3.26) into Eq. (3.25) then, subtracting from Eq. (3.24) finally, the result becomes to Eq. (3.10) in the text.

Appendix 2: Determination of Nonuniform Outflow Exponent (B)

To alternate the traditional relationship for the linear decreasing flow rate through the lateral line given by Eqs. (3.1a, 3.1b), to describe the variable flow rate along the pipeline Q(l), simple but sufficiently accurate exponential function can be used [33, 43, 48–51]

$$Q(l) = Q_I(1 - i)^{\beta} \tag{3.27}$$

If Eq. (3.27) is evaluated for the upper half segment of the pipeline ($i = 0.5$), then arranged with the expression for the flow rate entering over the upper half segment of the lateral, $Q_{0.5} = (kN/2)\overline{H}_{0.5}^{x}$ (k = coefficient of the emitter discharge) finally, the β can be evaluated from:

$$\beta = 1 - 3.322x \log\left(\frac{\overline{H}_{0.5}}{\overline{H}}\right) \tag{3.28}$$

where $\overline{H}_{0.5}$ = average pressure head along the upper half segment of the lateral.

Assuming the $\overline{H}_{0.5}$ is located in the middle of the length of the upper half segment of the lateral, and applying conservation of energy equation between the pipe segments l = L/4 and l = L, then the $\overline{H}_{0.5}$ can be expressed [21, 40]:

$$\overline{H}_{0.5} = H_N + H_{f(L)} + (p\overline{\Delta S}_{0.5} - \overline{H}_{f0.5}) \tag{3.29}$$

where H_N = downstream pressure head at the first emitter from the closed end of the lateral; $\overline{H}_{f0.5}$ = average head loss due to friction along the upper half segment of the lateral; and $\overline{\Delta S}_{0.5}$ = average head loss or gain due to elevation change at the end of the upper-quarter segment of the lateral, which is written by this form:

$$\overline{\Delta S}_{0.5} = \frac{3}{4}\Delta S \tag{3.30}$$

The H_N can be obtained for $i = 1$ from the pressure head profile given by Eq. (3.10) with taking into account the flow rate exponent, $m_\beta = 1.0 \times m$:

$$H_N = \overline{H} + \psi H_{VI} - \left(\frac{1}{m+2}\right)H_{f(L)} - \frac{p\Delta S}{2} \tag{3.31}$$

Average head loss due to friction between the pipe segments $l = 0$ and l = L/2 for the upper half segment of the lateral, $\overline{H}_{f0.5}$, is expressed by [21, 40]

$$\overline{H}_{f0.5} = \frac{2}{L}\int_{l=0}^{l=L/2} dS = \frac{2}{L}\int_{l=0}^{l=L/2} H_f(l)dl = \left(\frac{m + 0.5^{m+1}}{m+2}\right)H_{f(L)} \tag{3.32}$$

Substituting Eqs. (3.30)–(3.32) into Eq. (3.29), and dividing by $\overline{H}$ then, substituting into Eq. (3.28) finally, the result transforms to Eq. (3.18) in the text.

It should be noted that, an alternative form of Eq. (3.18) except for minor loss term, was presented in the recent papers [16, 21], on the basis of the average outlet flow approximation: $\left\{\overline{q} = (k/Ns)\left[\int_{l0}^{lQ_I} H^x(l)dl\right] \cong (k/Ns)\left[\int_{l0}^{lQ_I} H(l)dl\right]^x\right\}$ which directly referred to the paper [16].

However, according to the opinion of Wu [47] this approximation used is not possible in the mathematical consideration (the power of function x must be inside of the integration), and for different emitter models (except for laminar flow emitters, $x = 1.0$) it may produce results systematically in error. Further analysis and discussion for this approximation was presented in the recent papers [21, 31, 32, 40, 43, 50, 51]. As a matter of fact, an accurate average friction drop approach [42] is basically used along the background derivative procedure of the proposed analytical development (Appendices 1 and 2).

Appendix 3: Coefficient of Variation of Discharge (CVq) and Christiansen's Uniformity Coefficient (U_C)

1. The variance of pressure head profile between the pipe segments $l = 0$ and $l =$ L, can be

   expressed as [42]

$$H_{\text{var}} = \frac{1}{L}\int_{l=0}^{l=L} DEV_l^2 dl = \frac{1}{L}\int_{l=0}^{l=L} \left[H(l) - \overline{H}\right]^2 dl \tag{3.33}$$

where DEV_l = deviation between H(l) and $\overline{H}$ at any distance l from the pipeline inlet.

Subtracting $\overline{H}$ from Eq. (3.26) and into Eq. (3.33) then, solving the integration yields

$$\begin{aligned}
\text{H}_{\text{var}} = \Bigg(& H_{f(L)}^2 \frac{(m_\beta + 1)^2}{(m_\beta + 2)^2(2m_\beta + 3)} \\
& + H_{VI}^2\left\{\frac{2(m_\beta + 1)^2}{(m_\beta + 2)(2m_\beta + 3)}(1 - \psi)^2 + \left[\frac{1}{(m_\beta + 2)(m_\beta + 4)} - \frac{1}{3}\right]4(1 - \psi) + \frac{8}{15}\right\} \\
& + 2H_{f(L)}H_{VI}\left[\frac{(m_\beta + 1)^2}{(m_\beta + 2)^2(2m_\beta + 3)}(1 - \psi) - \frac{2}{3}\frac{(m_\beta + 1)}{(m_\beta + 2)(m_\beta + 4)}\right] \\
& + H_{VI}p\Delta S\left[\frac{(m_\beta + 1)}{(m_\beta + 2)(m_\beta + 3)}(1 - \psi) - \frac{1}{6}\right] \\
& + H_{f(L)}p\Delta S\frac{(m_\beta + 1)}{(m_\beta + 2)(m_\beta + 3)} + \frac{1}{12}(p\Delta S)^2\Bigg)
\end{aligned} \tag{3.34}$$

2. The relationship between CV_q and CV_H is given by [6, 7]

$$CV_q = xCV_H = \frac{x\sqrt{H_{\text{var}}}}{\overline{H}} \tag{3.35}$$

in which CV_q = the coefficient of variation of discharge is statistically defined as the standard deviation of discharge divided by the mean value and expressed in percent; and CV_H = the coefficient of variation of pressure head defined as standard deviation of pressure head divided by the average pressure head and expressed in percent The commonly used level of the CV_q varies for the intervals 3 and 5% [52–55].

3. For normal distribution, the Christiansen's uniformity coefficient U_C is related to CV_q by [42]:

$$U_C = 1 - 0.798CV_q \tag{3.36}$$

4. Substituting Eq. (3.34) for H_{var} into Eq. (3.35), then the CV_q into Eq. (3.36) finally, the U_C coefficient can be directly evaluated from Eq. (3.18), presented in the text.

References

1. R.W. Jeppson, *Analysis of Flow in Pipe Networks*, 5th edn. (Ann Arbor Science, Ann Arbor, Mich, 1982)
2. G. Yıldırım, O. Kose, New software (Multi-flowCAD) development for sub-main line hydraulic design: computational algorithm, computer visualization and implementation. Arab. J. Sci. Eng. **41**, 1551–1560 (2016)
3. G. Yıldırım, Computer-based analysis of hydraulic design variables for uniformly sloping microirrigation system laterals. J. Irrig. and Drain. Eng. ASCE **141**(7), 04014079 (2015)
4. G. Yıldırım, An assessment for hydraulically efficient design of uniformly sloping submain, lines. Irrig. Drain. **64**(2), 157–168 (2015)
5. G. Yıldırım, Total energy loss assessment for trickle lateral lines equipped with integrated in-line and on-line emitters. Irrig. Sci. **28**(4), 341–352 (2010)
6. G. Yıldırım, V.P. Singh, A MathCAD procedure for commercial pipeline hydraulic design considering local energy losses. Adv. Eng. Softw. **41**(2), 489–496 (2010)
7. G. Yıldırım, Computer-based analysis of explicit approximations to the implicit colebrook-white equation in turbulent flow friction factor calculation. Adv. Eng. Softw. **40**(11), 1183–1190 (2009)
8. G. Yıldırım, M. Özger, Neuro-fuzzy approach in estimating Hazen-Williams friction coefficient for small-diameter polyethylene pipes. Adv. Eng. Softw. **40**(8), 593–599 (2009)
9. G. Yıldırım, Simplified procedure for hydraulic design of small-diameter plastic pipes. Irrig. Drain. (ICID J.) **58**(3), 209–233 (2009)
10. M. Özger, G. Yıldırım, Determining turbulent flow friction coefficient using adaptive neuro-fuzzy computing technique. Adv. Eng. Softw. **40**(4), 281–287 (2009)

11. G. Yıldırım, Determining Operating Inlet Pressure Head Incorporating Uniformity Parameters for Multioutlet Plastic Pipelines. J. Irrig. Drain. Eng. ASCE **134**(3), 341–348 (2008)
12. G. Yıldırım, An assessment of hydraulic design of trickle laterals considering effect of minor losses. Irrig. Drain. (ICID J.) **56**(4), 399–421 (2007)
13. G. Yıldırım, Discussion of "statistical analysis of non-pressure-compensating and pressure-compensating drip emitters by H. Perea, J. Enciso-Medina, V. P. Singh, D. P. Dutta, and B. J. Lesikar . J. Irrig. and Drain. Eng. ASCE (2016). https://doi.org/10.1061/(ASCE)IR.1943-4774.0000644
14. H.M. Hathoot, A.I. Al-Amoud, F.S. Mohammad, Analysis and design of trickle irrigation laterals. J. Irrig. Drain. Eng. ASCE **119**(5), 756–767 (1993)
15. E.J. Scaloppi, R.G. Allen, Hydraulics of irrigation laterals: comparative analysis. J. Irrig. Drain. Eng. ASCE **119**(1), 91–115 (1993)
16. J.D. Valiantzas, Analytical approach for direct drip lateral hydraulic calculation. J. Irrig. Drain. Eng. ASCE **124**(6), 300–305 (1998)
17. J.D. Valiantzas, Continuous outflow variation along irrigation laterals: effect of the number of outlets. J. Irrig. Drain. Eng. ASCE **128**(1), 34–42 (2002)
18. P. Vallesquino, P.L. Luque-Escamilla, New algorithm for hydraulic calculation in irrigation laterals". J. Irrig. Drain. Eng. ASCE **127**(4), 254–260 (2001)
19. I.P. Wu, Energy gradient line approach for direct hydraulic calculation in drip irrigation design. Irrig. Sci. **13**, 21–29 (1992)
20. I.P. Wu, H.M. Gitlin, Energy gradient line for drip irrigation laterals. J. Irrig. Drain. Eng. ASCE **101**(4), 323–326 (1975)
21. G. Yıldırım, *Analytical relationships for designing multiple outlets pipelines* (ASCE, J. Irrig. Drain. Eng, 2006)
22. A.I. Al-Amoud, Significance of energy losses due to emitter connections in trickle irrigation lines. J. Agric. Eng. Res. **60**(1), 1–5 (1995)
23. V. Bagarello, V. Ferro, G. Provenzano, D. Pumo, Evaluating pressure losses in drip irrigation lines. J. Irrig. Drain. Eng., ASCE **123**(1), 1–7 (1997)
24. H.M. Hathoot, A.I. Al-Amoud, A.S. Al-Mesned, Design of trickle irrigation laterals considering emitter losses. ICID J. **49**(2), 1–14 (2000)
25. L. Juana, L. Rodriguez-Sinobas, A. Losada, Determining minor head losses in drip irrigation laterals. I: methodology. J. Irrig. Drain. Eng. ASCE 128(6), 376–384 (2002a)
26. L. Juana, L. Rodriguez-Sinobas, A. Losada, Determining minor head losses in drip irrigation laterals. II: experimental study and validation. J. Irrig. Drain. Eng. ASCE **128**(6), 385–396 (2002)
27. G. Provenzano, D. Pumo, Experimental analysis of local pressure losses for microirrigation laterals. J. Irrig. Drain. Eng. ASCE **130**(4), 318–324 (2004)
28. G. Provenzano, D. Pumo, P. Di Dio, Simplified procedure to evaluate head losses in drip irrigation laterals. J. Irrig. Drain. Eng. ASCE **131**(6), 525–532 (2005)
29. Y. Kang, S. Nishiyama, Analysis and design of microirrigation laterals. J. Irrig. Drain. Eng. ASCE **122**(2), 75–82 (1996)
30. P. Vallesquino, P.L. Luque-Escamilla, Equivalent friction factor method for hydraulic calculation in irrigation laterals. J. Irrig. Drain. Eng. ASCE **128**(5), 278–286 (2002)
31. G. Yıldırım, Discussion of 'Experimental analysis of local pressure losses for microirrigation laterals' by G. Provenzano and D. Pumo. J. Irrig. Drain. Eng. ASCE **132**(2), 189–193 (2006)
32. G. Yıldırım, N. Ağıralioğlu, Comparative analysis of hydraulic calculation methods in design of microirrigation laterals. J. Irrig. Drain. Eng. ASCE **130**(3), 201–217 (2004)

33. G. Yıldırım, Comparative analysis of hydraulic calculation methods in design oh micro-irrigation laterals and linear solution of hydraulic analysis of tapered laterals. PhD. thesis, Istanbul Technical University, Hydraulics Division, Istanbul, Turkey, 2007---\
34. V. Bagarello, D. Pumo, Lateral line hydraulics in drip irrigation systems, in Proceedings of the 16th European Regional Conference of ICID, ed. by L. Vermes, Budapest, Hungary (1992)
35. V. Bagarello, V. Ferro, G. Provenzano, D. Pumo, Experimental study on flow resistance law for small diameter plastic pipes. J. Irrig. Drain. Eng. ASCE 121(5), 313–316 (1995)
36. T.A. Howell, E.A. Hiler, Trickle irrigation lateral design. Trans. ASAE **17**(5), 902–908 (1974)
37. T.A. Howell, F.A. Barinas, Pressure losses across trickle irrigation fittings and emitters. Trans. ASAE **23**(4), 928–933 (1980)
38. J.E. Christiansen, Irrigation by sprinkling. Calif. Agric. Exp. Stn. Bull. **670**, (1942) (Univ. of California, Davis, California)
39. R.H. Cuenca, *Irrigation systems design: an engineering approach* (Prentice Hall, New Jersey, 1989)
40. G. Yıldırım, *Hydraulic analysis and direct design of multiple outlets pipelines laid on flat and sloping lands* (ASCE, J. Irrig. and Drain. Eng, 2006)
41. G.Z. Watters, J. Keller, Trickle irrigation tubing hydraulics. Presented at the Summer Meeting of ASAE (Utah State University, Logan, Utah, 1978), pp. 78–2015
42. H. Anyoji, I.P. Wu, Statistical approach for drip lateral design. Trans. ASAE **30**(1), 187–192 (1987)
43. G. Yıldırım, N. Ağıralioğlu, Linear solution for hydraulic analysis of tapered microirrigation laterals. J. Irrig. Drain. Eng. ASCE **130**(1), 78–87 (2004)
44. I.P. Wu, An assessment of hydraulic design of microirrigation systems. Agric. Water Manag. **32**, 275–284 (1997)
45. I.P. Wu, H.M. Gitlin, Drip irrigation design based on uniformity. Trans. ASAE **17**(3), 429–432 (1974)
46. I.P. Wu, R. Yue, Drip lateral design using energy gradient line approach. Trans. ASAE **36**(2), 389–394 (1993)
47. I.P. Wu, Personal communication, 2002
48. V.F. Bralts, I.P. Wu, H.M. Gitlin, Manufacturing variation and drip irrigation uniformity. Trans. ASAE **24**(1), 1234–1240 (1981)
49. V.F. Bralts, D.M. Edwards, I.P. Wu, Drip irrigation design and evaluation based on the statistical uniformity concept. Adv. Irrig. Acad. N.Y. **4**, 67–117 (1987)
50. G. Yıldırım, N. Ağıralioğlu, Closure to 'Comparative analysis of hydraulic calculation methods in design of microirrigation laterals' by G. Yıldırım and N. Ağıralioğlu. J. Irrig. Drain. Eng. ASCE **132**(1), 85–90 (2006)
51. G. Yıldırım, N. Ağıralioğlu, Closure to 'Linear solution for hydraulic analysis of tapered microirrigation laterals' by G. Yıldırım and N. Ağıralioğlu. J. Irrig. Drain. Eng. ASCE **131**(5), 490–491 (2005)
52. V. Ravikumar, C.R. Ranganathan, S. Santhana Bosu, Analytical equation for variation of discharge in drip irrigation laterals. J. Irrig. Drain. Eng. ASCE 129(4), 295–298 (2003)
53. R.D. von Bernuth, Simple and accurate friction loss equation for plastic pipe. J. Irrig. Drain. Eng. ASCE **116**(2), 294–298 (1990)
54. R.D. von Bernuth, T. Wilson, Friction factors for small diameter plastic pipes. J. Hydraulic Eng. ASCE 115(2): 183–192 (1989)
55. G. Yıldırım, *Computer-aided design of micro-irrigation laterals by using LatCAD* (Thesis, Istanbul Technical University, Hydraulics Division, Istanbul, Turkey, MSc, 2001)

Analytical Direct Hydraulic Design Procedure for Determining Pipeline Geometric Characteristics (Pipe Diameter and Length) for Flat and Uniformly Sloping Multioutlets Submain Lines: Methodology, Implementation, Comparative Analysis

4

Introduction

Hydraulic design of a multiple outlets pipeline is one of the important factors in the ultimate success or failure of irrigation systems. A multiple outlets pipeline is a hydraulic structure whose design is limited by the operating inlet pressure head, and by water application uniformity that is affected by the total energy loss along the pipeline, the field topography, as well as the outlet hydraulic characteristics.

In the past, numerous researchers have worked on the hydraulic analysis of multiple outlets pipeline networks [1–12]. As a matter of fact, a significant amount of these hydraulic analysis are based on the conventional design procedure in which the pipe size and length are known a priori; then the remaining design variables such as operating inlet pressure head, water application uniformity and total energy loss are determined depending on these known pipe geometric characteristics. However, designers are often faced with the problem of selecting the pipe size and length for a given desired uniformity level and given design slope range, in order to keep the design operating inlet pressure head and the allowed limit of total energy loss along the pipeline. The author [12] developed a design curve for a trickle lateral with turbulent flow emitters. Using the curve, a lateral length or diameter with operating inlet pressure head can be designed when the required average emitter discharge, the required uniformity of water application, and other conditions are given.

To analyze and design a trickle lateral, [1–111 presented a stepwise (SBS) method which neglects the emitter local losses, and developed the design curves similar to that presented [12]. The curves are essentially based on the conventional design procedure aforementioned above, and can be only used for the same example data presented in their paper. In the following work [12], the SBS method was modified to suit sprinkler lateral design considering the head loss in sprinkler risers as well as the lateral fitting losses, and

G. Yıldırım, *Hydraulic Principles and Design Concepts for Submain Units with Multiple Outlet Pipelines*, Synthesis Lectures on Engineering, Science, and Technology,
https://doi.org/10.1007/978-3-031-32495-6_4

a flowchart for a computer program employing this technique was also presented so as to design the lateral diameter [13].

Recently, a revised SBS (R-SBS) method [13] was also developed for a trickle lateral which considers the local losses due to emitter connections, and a flowchart for a computer program for designing either the diameter or the length of lateral was also provided. In the analysis it was reported that, in some design cases, neglecting emitter connection losses may lead to designs in error by 30% [13–16].

In the last decade, the finite element method (FEM) was applied for designing single and paired microirrigation laterals on both flat and sloped fields [17–19]. In the analysis it was shown that, the lateral length for the required uniformity of water application may have three solutions and the diameter may have two solutions when a single lateral is laid on a downhill sloped field, or paired laterals are laid on a sloped field. Both the R-SBS and the FEM methods have the highest accuracy because only the basic equations of the hydraulics of steady pipe flow were used; however, both the procedures require large amounts of programming time and complex computer operations.

The author [20, 21] presented an analytical procedure for trickle laterals, and also proposed an explicit equation for direct design of the length of a horizontal lateral. However, several relationships presented in this procedure were discussed earlier [22–24] based on the noteworthy opinion [58], related to the basic derivative procedure in the background of some analytical expressions. The authors [25] have proposed a quadratic analytical equation to direct design of a trickle lateral diameter. In the paper it was reported that, two realistic solutions of diameters exist for the down slope of 5% and above and, both the diameters are imaginary indicating realistic solutions do not exist for an up slope of 3% and above.

In a recent study, the state-of-the art review of some recent studies for microirrigation lateral design [11, 24, 28] was presented. In the analysis, a detailed comparison test was applied to several design examples to cover various combinations of design parameters and varying emitter discharge exponents, then the dimensionless design charts for different ranges of lateral diameter and length and for various uniform line slope cases, are also presented.

This chapter presents an analytical procedure allowing the hydraulic analysis and/or the direct solution of the design problem of a single multiple outlets pipeline. In the first stage of this analysis, analytical expressions taking into consideration the influence of local energy loss [1, 14–16, 29–39] are derived, on the basis of the conventional design procedure presented in the earlier work [40]; then in the following stage, for any desired uniformity level and given design slope with remaining known parameters, analytical formulations allowing direct design of the pipe size and length are respectively deduced.

The proposed equations are suitable for designing laterals and manifolds in both trickle and sprinkler irrigation systems, and can be applied for various types of outlet, different flow regimes and uniform line slope ranges [14]. The comparison test between the proposed design procedure with the R-SBS method [11–13], for different slope combinations indicated that, the former can be considered sufficiently accurate for various design cases in both trickle and sprinkler systems.

Hydraulic Analysis

Despite the large variety of irrigation laterals, two main conditions are well characterized in terms of multiple outlets pipelines [41–61]. In the first one, the discharge at each equally spaced outlet is assumed to be constant, and the resulting outflow along the lateral should be as uniform as possible. In the second condition, the outflow discharge is designed to be variable along laterals. These conditions are usual in most sprinkler and trickle irrigation laterals, manifolds and gated pipes for surface irrigation, especially those with a large number of outlets [30]. Analytical development having regard to both the conditions for designing multiple outlets pipelines was earlier presented [40].

Systematic hydraulic analysis of both sprinkler and trickle laterals was presented in the existing literature [11–13, 62–69]. Comparative analysis of sprinkler and trickle laterals [12, 13] indicates that the design problems of both laterals can be generally solved by the four basic equations of the steady pipe flow (the outlet discharge-pressure head relationship, the continuity equation, the conservation of energy equation with the Darcy-Weisbach friction formula and the conservation of momentum equation); and the same uniformity criteria -the Christiansen's uniformity coefficient, U_C [41, 47] and variation of the outlet operating pressure head, δ_H-can be directly used in design of both laterals [6]. Flow conditions and corresponding friction coefficients for both laterals will be evaluated in the following section.

Methodology

Energy Relations

Flow in a multiple outlets pipeline is considered to be steady, spatially varied flow with decreasing discharge along the direction of flow. Assuming that outlets along the pipeline have uniform outflow $\overline{q}$, then the total flow rate that passes through the pipe section located at any distance l from the pipeline inlet, Q(l), is commonly given as [9, 51, 70]

$$Q(l) = \left(N - \frac{l}{s}\right)\overline{q} = Q_I(1 - i) \tag{4.1}$$

where Q(l) = the discharge at a section of length l measured from the pipe inlet (m^3s^{-1}); $Q_I = N\overline{q}$ = total inflow rate for the entire length of the pipe (m^3s^{-1}); N = total number of equally spaced operating outlets for the entire pipeline; l = original distance which is measured from the pipeline inlet (m); s = outlet spacing (m); and i = dimensionless variable of distance $i = l/L$, which is the ratio of the length to the total length (m/m).

The concept of energy grade line or the friction drop ratio approach [9] is valuable in hydraulic analyzing of steady pipe flow problems. The actual energy grade line in a pipeline with multiple outlets will not be a straight line but a curve of exponential type and, gradually decreases in the direction of flow, therefore, the nonuniform outflow profile is actually evident along the pipeline.

The slope of the energy grade line at any distance l from the pipeline inlet, $S_f(l)$, is generally expressed by the accurate Darcy-Weisbach formula [40]:

$$S_f(l) = \frac{dH_f(l)}{dl} = \frac{f}{D} H_V(l) = \frac{C}{D^n} Q^m(l) \tag{4.2a}$$

$$f = c_1 [R_I]^{m-2} \tag{4.2b}$$

$$C = \frac{8c_1(\upsilon/4)^{2-m}}{g\pi^m} \tag{4.2c}$$

where $S_f(l)$ = friction slope or the slope of the energy grade line at any distance l measured from the pipeline inlet expressed as the energy loss per unit length of the pipeline-per unit weight of fluid (m/m); $H_f(l)$ = friction head loss per unit weight of fluid at any distance l (m); $H_V(l)$ = velocity head for the pipe section located at any distance l (m); D = internal diameter of the pipeline (m); f = friction coefficient for the Darcy-Weisbach formula assuming to be constant along the pipeline; C = combined friction coefficient coupled with f, for the Darcy-Weisbach formula; c_1 = constant for the particular friction coefficient formula used; $R_I = 4Q_I/\pi D\nu$ = Reynolds number at the inlet of the pipeline; m and n = velocity (or flow rate) and diameter exponents, respectively. Generally n is approximately equal to m + 3 [30]; ν = kinematic viscosity of water (m^2s^{-1}); and g = acceleration due to gravity (ms^{-2}).

Flow in laterals is usually turbulent, sometimes fully turbulent flow exists at the upstream end of the lateral and, flow becomes laminar at the downstream reach where the flow velocity decreases to zero. The transition flow approximately occurs for the interval, $2{,}000 < R \leq 3{,}000$ and the flow characteristics are indeterminate for the transition zone [49].

Sprinkler laterals are either plastic or aluminum pipes, whereas trickle laterals are generally made of smooth plastic material. For plastic pipes the friction coefficient f can be evaluated from Eq. (4.2b), and flow conditions are the following: For laminar flow regime ($R < 2{,}000$; R = Reynolds number for pipe flow), m = 1.0, n = 4.0, c_1 = 64 and $C = 4.195 \times 10^{-6}$; for turbulent flow regime in a smooth pipe ($3{,}000 < R \leq 10^5$), m = 1.75, n = 4.75, c_1 = 0.316 (Blasius equation) and $C = 7.792 \times 10^{-4}$, and finally for fully turbulent flow regime ($10^5 < R < 10^7$), m = 1.828, n = 4.828, c_1 = 0.130 and $C = 9.589 \times 10^{-4}$, are taken [59].

In aluminum pipes, the absolute roughness ranges between 0.1 and 0.3 mm [46]. For laminar flow, the f is computed as mentioned above; and for turbulent flow case, one of the explicit forms of the Colebrook-White equation [42, 52], can be directly used.

Integrating Eq. (4.2a) with Eq. (4.1) between the limits 0 and l yields [5, 9]

$$R_i = \frac{H_f(l)}{H_{f(L)}} = \left[1 - (1 - i)^{m+1}\right] \tag{4.3a}$$

$$H_{f(L)} = \overline{F} H_{f0} = \overline{F}\left(C \frac{Q_I^m}{D^n} L\right) \tag{4.3b}$$

where R_i = friction drop ratio which is expressed as the percentage of the friction head loss at any distance l to the total friction drop at the downstream closed end of the pipeline (m/m); $H_{f(L)}$ = total friction drop at the closed end of the pipeline (m); $\overline{F}$ = friction correction factor; H_{f0} = total friction drop in a similar pipeline without outlets conveying the total flow rate along its entire length (m); and L = full length of the pipeline (m).

For the traditional approach that takes into account the continuous and uniform outflow along the pipeline, $\overline{F}$ takes the form:

$$\overline{F} = \frac{1}{(m+1)} \tag{4.4}$$

To alternate Eq. (4.4) the Christiansen's correction factor [7], $\overline{F}_C$, based on uniform and discrete outflow from finite number of operating outlets, for the case in which the first outlet is located in a full outlet spacing from the pipeline inlet, is given by the following

$$\overline{F}_C = \frac{1}{(m+1)} + \frac{1}{2N} + \frac{\sqrt{m-1}}{6N^2} \tag{4.5a}$$

If the first outlet is half outlet spacing from the pipeline inlet, F_C is given by [9]:

$$\overline{F}_C = \frac{2N}{(2N-1)}\left(\frac{1}{m+1} + \frac{\sqrt{m-1}}{6N^2}\right) \tag{4.5b}$$

It is worthy to note, the $\overline{F}_C$ by Eq. (4.5) tends to give smaller values with increasing the number of outlets N; and the $\overline{F}_C$ by Eq. (4.5a) systematically yields bigger values than those of Eq. (4.5b). The difference between Eqs. (4.5a) and (4.5b) is negligible as the number of outlets increases. For instance (for m = 1.75), for a relatively large value of N = 150 the difference is 0.6%, whereas for an extreme small value of N = 5, the difference is 12.5%, approximately.

To determine the total amount of actual energy losses along a multiple outlets pipeline, head losses through local singularities at the outlet connections must be additionally taken into account. These minor losses, $H_{s(L)}$, are usually calculated as if the pipe length were increased by the so-called equivalent length, l_e, i.e., a length of the same uniform pipe that would have the same head loss [30–32].

Total energy losses along the lateral, $H_{T(L)}$, includes friction and local head losses can be formulated by the following expression:

$$H_{T(L)} = H_{f(L)} + H_{s(L)} = F_S H_{f(L)} = \overline{F} F_S \left(C \frac{Q_I^m}{D^n} L \right) \tag{4.6a}$$

where

$$F_S = \left(1 + \frac{l_e}{s} \right) \tag{4.6b}$$

in which F_S = amplification factor to be applied to the friction losses in the uniform pipe sections, in order to take into account the effect of local losses; and l_e = equivalent length of a minor singularity (m) [31, 32].

The recent studies of local losses in trickle laterals [1, 29, 33–36] reported that, the local losses due to the emitter insertions increases with decreasing the lateral diameter and with increasing the lateral length in relation to the high number of emitters located along the pipeline. Note that for trickle laterals, the values of l_e for different type of emitters, characterized by different flow regimes, can be calculated on the basis of experiments recently carried out [31].

For sprinkler laterals, the amount of local losses can be evaluated considering both the head loss due to a coupler which connects two lengths of plain pipe sections, and the head losses between the upstream and downstream points of a sprinkler riser and along its height. Note that in the present analysis, the local losses due to couplers are only taken into account whereas relatively minor local losses due to sprinkler risers are considered negligible.

For plastic pipes the local loss in a coupler could be provided as equivalent length of the pipe [12, 50], and for aluminum pipes it can be evaluated by increasing the friction coefficient f by a fraction Δ $[(1 + \Delta) f]$. For couplers spaced 9 m, the energy loss can be determined by increasing the f by 11% ($\Delta = 0.11$). Decreasing the spacing to 6 m increases the percentage 17% ($\Delta = 0.17$), and increasing the spacing to 12 m decreases the percentage to 8% ($\Delta = 0.08$) [12, 46].

The pressure head change from the pipeline inlet to a given length l, is a linear combination of the pressure head changes due to friction and uniform slope along the pipeline. Applying the conservation of energy equation between the pipe segments $l = 0$ and $l = l$, then the pressure head profile, $H(l)$, can be determined using the similar procedure presented in the earlier work [40]:

$$H(l) = \overline{H} + H_{VI}\left[1 - (1 - i)^2\right] + H_{f(L)}\left[F_S\left[(1 - i)^{m_\beta + 1} - 1\right] + \frac{m_\beta + 1}{m_\beta + 2}\right] - p\Delta S\left(i - \frac{1}{2}\right) \tag{4.7}$$

where H(l) = pressure head at a given section of length, l from the pipeline inlet (m); $\bar{H}$= average pressure head for the entire of the pipeline (m); $H_{VI} = 8(N\bar{q})^2/\pi^2 g D^4$= velocity head at the pipeline inlet (m); $m_\beta = \beta\times$ m: improved value of the velocity exponent m, given by Eq. (4.8); $\Delta S = S_0 L$ = total head loss or gain due to change of elevation at the closed end of the pipeline (m); and S_0 = line slope assuming to be uniform along the pipeline (m/m). Noting that in this procedure, the slope of a downhill pipeline is negative (p = --1), and the slope of an uphill pipeline is positive (p = 1).

According to this analysis presented above, the energy grade line based on the nonuniform outflow concept is determined by Eq. (4.7) in which the velocity exponent m should be replaced by its improved value $m_\beta = \beta \times$ m then, the nonuniform outflow exponent, β, can be approximated by the following relationship (*Appendix II*):

$$\beta = 1 - 3.322x\log\left[1 + \frac{H_{VI}}{\overline{H}} + \frac{H_{f(L)}}{\overline{H}}\left[\left(\frac{m+3-0.5^{m+1}}{m+2}\right) - F_S\right] + \frac{1}{4\overline{H}}(p\Delta S)\right] \tag{4.8}$$

in which β = nonuniform outflow exponent then, the outflow from each multiple outlet will not be uniform along the pipeline.

Uniformity Criteria

The hydraulic design procedure for a pipeline with multiple outlets usually establishes the maximum difference in the extreme outlet operating pressure heads along the pipeline is equal to the maximum allowed difference, ΔH. The uniformity criterion applied for the design of multiple outlets pipelines is that the maximum allowed difference in the outlet operating pressure head along the pipeline, ΔH, is less than a percentage of the average pressure head, δ_H:

$$\delta_H = \frac{\Delta H}{\overline{H}} = \frac{(H_{\max} - H_{\min})}{\overline{H}} \tag{4.9}$$

where δ_H = percentage of the maximum allowable difference in the outlet operating pressure heads to the average outlet pressure head (m/m); ΔH = maximum allowable difference in the outlet operating pressure head along the pipeline (m); and $H_{max}-H_{min}$ = maximum difference in the extreme outlet operating pressure heads along the pipeline (m). The usual criterion applied for the design of laterals is that the difference in outlet discharge along a single lateral is less than ±10%. This corresponds to a maximum difference in outlet operating pressure head of ±20% [43].

Considering multiple outlets along the pipeline being equally spaced then, the inlet spacing s_I, from the pipeline inlet to the first outlet is expressed as the some fraction, μ ($0 \le \mu \le$ 1.0.01. 0.0) of the common outlet spacing s, s_I = μ s, and it usually takes 0.5 or 1.0 for multiple outlets pipelines.

In downhill lines, the location of the maximum and the minimum pressure heads is variable and depends on the relationship between the total energy gain or loss due to uniform slope and the total head losses due to pipe friction. The location of the maximum pressure head is found either at the first outlet near the pipeline inlet (i_{max} = μ s/L) or at the first outlet from the downstream closed end point (i_{max} = 1.0). In downhill slope, the current design procedure assumes the most usual case in which the maximum operating outlet pressure head occurs at the first outlet near the downstream closed end of the pipeline

$$i_{\max} = \frac{l_{\max}}{L} = 1 \tag{4.10a}$$

The location of the minimum pressure head is determined where the first derivative of the pressure head distribution given by Eq. (4.7) is equal to zero. For simplification of this derivative procedure, the velocity head term can be omitted in this equation, then the location i_{min} where the minimum pressure head occurs, can be approximated by

$$i_{\min} = \frac{l_{\min}}{L} = 1 - \left[\frac{(-p\Delta S)}{F_S H_{f(L)}(m_\beta + 1)}\right]^{1/m_\beta} \tag{4.10b}$$

Using Eq. (4.10b) for i_{min} with substituting Eq. (4.10a) into Eq. (4.7) and, the result into Eq. (4.9), then the δ_H is expressed by

$$\delta_H = \frac{\left[H_{VI}(1 - i_{\min})^2 - F_S H_{f(L)}(1 - i_{\min})^{m_\beta+1} - p\Delta S(1 - i_{\min})\right]}{\overline{H}} \tag{4.11a}$$

In uphill and zero slope conditions, the maximum pressure head is always located at the first outlet from the pipeline inlet (i_{max} = μ s/L), whereas the minimum pressure head is always found at the first outlet from the downstream closed end of the pipeline (i_{min} = 1.0). Using the expressions for i_{max} and i_{min} in Eq. (4.7) and the result into Eq. (4.9), finally the δ_H can be similarly obtained as follows

$$\delta_H = \frac{\left(F_S H_{f(L)}\theta^{m_\beta+1} - H_{VI}\theta^2 + p\Delta S\theta\right)}{\overline{H}} \tag{4.11b}$$

where $\theta = L_1/L$ = length ratio which is the percentage of the distance between the first and last outlets to the full length of the pipeline (m/m); $L_1 = (N - 1)s$ = length of the distance between the first and last outlets except for the inlet spacing, s_I (m); and $L = (N + \mu - 1)s$ = full length of the pipeline as defined in the previous section (m).

The variance of pressure head profile, H_{var} between the pipe segments $l = 0$ and $l =$ L, is expressed by [5]:

$$H_{\text{var}} = \frac{1}{L}\int_{l=0}^{l=L} DEV_l^2 dl = \frac{1}{L}\int_{l=0}^{l=L} \left[H(l) - \overline{H}\right]^2 dl \tag{4.12a}$$

where DEV_l = deviation between $H(l)$ and $\bar{H}$ at any distance l from the pipeline inlet (m).

Subtracting $\bar{H}$ from Eq. (4.7) and into Eq. (4.12) then, solving the integration yields:

$$\begin{aligned}\mathrm{H_{var}} &= H_{f(L)}^2\left[2F_S^2\frac{(m_\beta+1)^2}{(m_\beta+2)(2m_\beta+3)}+(1-2F_S)\frac{(m_\beta+1)^2}{(m_\beta+2)^2}\right]\\ &+\frac{4}{3}\frac{(m_\beta+1)}{(m_\beta+2)}\left[1-F_S\frac{(m_\beta+5)}{(m_\beta+4)}\right]H_{VI}H_{f(L)}\\ &+\frac{(m_\beta+1)}{(m_\beta+2)(m_\beta+3)}F_S H_{f(L)}(p\Delta S)+\frac{8}{15}H_{VI}^2-\frac{1}{6}(p\Delta S)H_{VI}+\frac{1}{12}(p\Delta S)^2\end{aligned} \tag{4.12b}$$

For a normal distribution case, the H_{var} is related to the Christiansen's uniformity coefficient, U_C by the following expression [12].

$$H_{\mathrm{var}} = \left[\frac{(1-U_C)\overline{H}}{0.798x}\right]^2 \tag{4.12c}$$

where x = outlet discharge exponent characterizing the flow regime and the outlet type.

Design Formulations

To this point of the present analysis, analytical expressions are based on the conventional design procedure in which the pipe size and length are known a priori, and they are derived under the previous work [46]. In the following section, the practical equations allowing the direct design solution of a multiple outlets pipeline–design of pipe size and length–suitable for different slope ranges and different flow regimes, will be deduced.

Design of Pipe Diameter

For simplification of the design procedure, the kinetic head terms can be neglected in Eq. (4.12b) and finally, from the equality of the right hand side of Eqs. (4.12b) and (4.12c) then, the following expression can be written:

$$\begin{aligned}&H_{f(L)}^2\left[2F_S^2\frac{(m_\beta+1)^2}{(m_\beta+2)(2m_\beta+3)}+(1-2F_S)\frac{(m_\beta+1)^2}{(m_\beta+2)^2}\right]\\ &+\frac{(m_\beta+1)}{(m_\beta+2)(m_\beta+3)}F_S H_{f(L)}(p\Delta S)+\frac{1}{12}(p\Delta S)^2\\ &-\left[\frac{(1-U_C)\overline{H}}{0.798x}\right]^2=0\end{aligned} \tag{4.13}$$

Rearranging of Eq. (4.13) for the D, finally it transforms to the following general equation in a quadratic form:

$$X\left(\frac{1}{D^{m+3}}\right)^2 + Y\left(\frac{1}{D^{m+3}}\right) + Z = 0 \tag{4.14a}$$

where X, Y and Z = constants of the quadratic equation as given by the following expressions:

$$\mathrm{X} = C^2\overline{F}_C^2 Q_I^{2m} L^2 K_1 \tag{4.14b}$$

$$\mathrm{Y} = C F_S \overline{F}_C Q_I^m (p\Delta S) L K_2 \tag{4.14c}$$

$$\mathrm{Z} = \frac{(p\Delta S)^2}{12} - \left[\frac{(1 - U_C)\overline{H}}{0.798x}\right]^2 \tag{4.14d}$$

where K_1 and K_2 = constants, are given as follows:

$$\mathrm{K}_1 = 2F_S^2 \frac{(m_\beta + 1)^2}{(m_\beta + 2)(2m_\beta + 3)} + (1 - 2F_S)\frac{(m_\beta + 1)^2}{(m_\beta + 2)^2} \tag{4.14e}$$

$$\mathrm{K}_2 = \frac{(m_\beta + 1)}{(m_\beta + 2)(m_\beta + 3)} \tag{4.14f}$$

As indicated in Eq. (4.14) it is worthy to note, for a given design slope S_0, for the desired water application uniformity, U_C and, for the design length of the pipe, L (m); the design pipe diameter, D (m) can be directly determined for initially known values of the following design variables: the correction factor, $\overline{F}_c$ from Eq. (4.5); the enlargement factor, F_S from Eq. (4.6b); the full length of the pipeline, L (m); the change of elevation due to given design slope, $\Delta S = S_0 L$ (m); the inlet flow rate through the pipeline, Q_I ($m^3 s^{-1}$); the average pressure head of the outlet, (m), the outlet discharge exponent, x; and the improved velocity exponent for nonuniform outflow, m_β from Eq. (4.8).

Solving Eq. (4.14) would yield two diameter values. It is worthy to note, both diameter values for a fixed value of the uniformity coefficient, U_C has already been reported in a recent study [19]. Among of two diameters, the minimum positive diameter can be taken as the design diameter and the second one diameter available in the market can be selected for adoption [25].

Design of Pipe Length

Rearranging of Eq. (4.13) for L then, the design length of the pipe can be determined by the following general equation:

$$\left[C^2\overline{F}^2 D^{-2(m+3)} K_1\left(\frac{\overline{q}}{s}\right)^{2m}\right]L^{2(m+1)} + \left[CF_S\overline{F}pS_0D^{-(m+3)}K_2\left(\frac{\overline{q}}{s}\right)^{m}\right]L^{m+2} + \frac{(pS_0)^2}{12}L^2 - \left[\frac{(1-U_C)\overline{H}}{0.798x}\right]^2 = 0 \quad (4.15a)$$

Noting that Eq. (4.15a) is deduced assuming by the first outlet from the pipeline inlet is full outlet spacing; however, for different position of the first outlet it has enough accuracy, as well. Unfortunately, the solution of Eq. (4.15a) for L requires a trial-and-error procedure depending on the initially known values of the main design parameters S_0, U_C and, D with remaining known parameters, such as average outflow, $\overline{q}$ (m^3s^{-1}), outlet spacing, s (m), and the others mentioned in Eq. (4.14). However, most of software packages in the generic mathematics (i.e., *Matlab*, *Excel* Equation-Solver or *MathCAD* Equation-Solver) can easily solve it [44].

As a special case, for horizontal pipelines the design length can be directly computed from:

$$L = \frac{\left[\frac{(1-U_C)\overline{H}}{0.798x}\right]^{1/(m+1)}}{\left[C^2\overline{F}^2 D^{-2(m+3)} K_1\left(\frac{\overline{q}}{s}\right)^{2m}\right]^{1/2(m+1)}} \quad (4.15b)$$

Design Procedure

Design of a multiple outlets pipeline often requires the determination of the pipe diameter or length for the design slope, S_0 and for the desired water application uniformity level, U_C. In this procedure, the design solution can be obtained in two computation stages. In the first stage, the values of the main design parameters can be found assuming turbulent flow and the uniform outflow occurred along the pipeline; then in the second improved stage, using Eqs. (4.14) or (4.15) the values of design parameters can be determined depending on the real flow regime occurred at the pipeline inlet (either fully-turbulent and turbulent) and with taking into consideration the nonuniform outflow profile. It should be noted, in most design cases the first stage of this procedure produces sufficiently accurate results with a little deviation; however, if a greater precision is required the second improved stage can then be applied. More clarification for the design procedure was earlier presented [40].

Application and Verification

To illustrate the practical importance of the proposed analytical procedure for determining pipe size and length, the following design examples for sprinkler and trickle irrigation systems, will be presented.

Example 1: Design of Sprinkler Lateral Diameter

In order to demonstrate the application of the proposed Eq. (4.14) the numerical example similar to that presented earlier [12], is used for the sake of comparison.

It is required to design the diameters D_1, D_2 and D_3 for horizontal sprinkler lateral, for the desired uniformity levels $U_{C1} = 76.7\%$, $U_{C2} = 95.8\%$ and $U_{C3} = 98.9\%$, and for the equivalent lengths for plastic couplers, $l_{e1} = 1.83$ m, $l_{e2} = 2.44$ m and $l_{e3} = 3.35$ m, respectively. Consider a sprinkler lateral is of PVC material (x = 0.5) and use the following data: total number of sprinklers, N = 33 which are equally spaced at s = 12. 0 m; the inlet spacing ratio from the main pipe, $\mu = s_I/s = 1.0$. Irrigation water is at 20 °C (kinematic viscosity, $\nu = 1.01 \times 10^{-6}$ m^2s^{-1}); and acceleration due to gravity, g = 9.807 ms^{-2}. The required average sprinkler discharge and corresponding outlet pressure head are, $\overline{q} = 0.315$ ls^{-1} and = 31.61 m, respectively.

The amplification factors related with the corresponding equivalent lengths are computed from Eq. (4.6b) as follows: $F_{S1} = 1.153$, $F_{S2} = 1.203$ and $F_{S3} = 1.279$, respectively. For the desired uniformity levels $U_{C1} = 76.7\%$, $U_{C2} = 95.8\%$ and $U_{C3} = 98.9\%$, Eq. (4.14) yields the diameters $D_1 = 51.3$ mm (2 in.), $D_2 = 75.3$ mm (3 in.), and $D_3 = 101.6$ mm (4 in.), respectively. It is worthy to note, the diameter values increase with increasing the desired uniformity levels, as well as with increasing the equivalent length values.

In order to evaluate the influence of coupler energy loss on the pipe diameter, D (mm), operating inlet pressure head, H_I (m), and total percentage energy losses, δ_H (%), these design parameters (D, H_I and δ_H) are computed for both the following design cases, design case (**a**): **without coupler energy loss**, and design case (**b**): **including coupler energy loss**, and the results are shown in Table 4.1. From the table, for the design case (**b**), for three desired uniformity levels, the pipe diameters increase 4.1%, 6.5% and 8.4% respectively, with respect to the design case (**a**). On the other hand, values of the H_I decrease 10.3%, 4.5%, and 1.8%, respectively, as well as the δ_H decreases due to pipe size increasing, 1.7%, 6.8%, and 10.6% for three uniformity levels, respectively. In other words, the percentage decreasing in H_I decreases with increasing D whereas the percentage decreasing in δ_H increases with increasing D.

The present example was earlier solved [12] for the case of an aluminum sprinkler lateral ($\varepsilon = 0.1$ mm, ε = absolute roughness of the pipe). Note that in their computation, the friction coefficient f is multiplied by 1.08 ($\Delta = 0.08$) to account for losses in couplers 12.0 m apart. For the sake of comparison, for three desired uniformity levels, the design parameters, D (mm), H_I (m), and δ_H (%), for PVC pipe obtained from the proposed

Table 4.1 Comparison of values of D, H_I and δ_H obtained with the proposed design procedure for the design of a horizontal sprinkler lateral (Example 1), and percentage differences between the design cases: case (a): without coupler energy loss ($F_S = 1.0$), and case (b): including coupler energy loss ($F_S = 1.153$, 1.203 and 1.279)

Design parameters	Design case (a): without coupler energy loss			Design case (b): including coupler energy loss			Differences between the design cases (a) and (b)		
	$F_S = 1.0$			$F_S = 1.153$	$F_S = 1.203$	$F_S = 1.279$			
	$U_C = 76.7\%$ (1)	$U_C = 95.8\%$ (2)	$U_C = 98.9\%$ (3)	$U_C = 76.7\%$ (1)	$U_C = 95.8\%$ (2)	$U_C = 98.9\%$ (3)	(1)	(2)	(3)
D (mm) Eq. (4.14)	49.3	70.7	93.7	51.3	75.3	101.6	Increasing in D (%)		
							4.1	6.5	8.4
H_I (m) Eq. (4.7)	78.6	40.1	33.8	70.5	38.3	33.2	Decreasing in H_I (%)		
							10.3	4.5	1.8
δ_H (%) Eq. (4.11b)	181.6	32.5	8.5	178.5	30.3	7.6	Decreasing in δ_H (%)		
							1.7	6.8	10.6

method and for aluminum pipe according to the revised Step-by-Step (R-SBS) method [46], are comparatively presented in Table 4.2. As indicated in the table, for the pipe diameter and for three desired uniformity levels, both the methods give similar results. However, for three uniformity levels, for the design of aluminum pipe, H_I and δ_H yield higher values with respect to the design of PVC pipe. It is considered from this table, the percentage increasing in H_I and δ_H decreases with increasing the desired uniformity levels.

It should also be noted, the same example for the design case of an aluminum sprinkler lateral was examined in the earlier work [46]. A comparison test for the three design parameters (D, H_I and δ_H) shows that, the results of the conventional design procedure [40] and those of the R-SBS numerical method are also in a good agreement.

Example 2: Design of Microirrigation Lateral Diameter

It is required to design the diameter, D, of a horizontal PE microirrigation lateral, as well as the operating pressure head at the pipe inlet, H_I, and the amount of total energy losses, $H_{T(L)}$, considering both the design cases (a) without local energy loss (F_S = 1.0) for U_C = 95.3% and, (b) including local energy loss (F_S = 1.26) for U_C = 98.1% using the following data very similar to the one presented [13]: The local energy loss of an orifice vortex on-line emitter model is characterized by l_e = 0.26 m [17]. The number of emitters, N = 101; the spacing between successive emitters, s = 1.0 m with the inlet spacing ratio, $\mu = s_I/s = 1.0$; the emitter discharge exponent, x = 0.2; the average emitter discharge $\overline{q}$ = 4.0 lph (1.111×10^{-6} m^3s^{-1}) and corresponding pressure head, $\bar{H}$ = 9.6 m, respectively.

On the basis of the algorithm discussed, the design case (a) gives the results: D = 10.3 mm, H_I = 16.75 m, and $H_{T(L)}$ = 9.75 m, whereas the design case (b) yields D = 13.4 mm, H_I = 11.65 m, and $H_{T(L)}$ = 3.52 m, respectively. It is evident from these results, considering local energy losses has the effect of increasing the pipe diameter 30.1%, decreasing the operating inlet pressure head 30.4%, as well as decreasing the total energy loss due to pipe size increasing 63.9%, respectively (Please also see, Chap. 3, pages: 82–83, "*Example 1: Determination of Lateral Diameter Using Dimensionless Design Curves*").

For this example, the R-SBS numerical method [13] yields similar results for the pipe size; for the case of without emitter losses, D = 10 mm and for the case of including emitter losses, D = 13 mm for small, medium and large size of emitter barb, as well. It means that, considering the emitter local losses increases 30% the pipe diameter with respect to the case of neglecting emitter local losses, according to the R-SBS method [13].

In order to better analyze the results of the proposed algorithm for various slope combinations (S_0 = 0.0, ∓ 0.005, ∓ 0.01, ∓ 0.02, ∓ 0.03, ∓ 0.04, and ∓ 0.05), the above design problem is solved with the same data, as well. For comparative purpose, the complete results for the design parameters obtained from the proposed method and the R-SBS method [13] with the small size of emitter barb are presented in Table 4.3. As shown in the table for both the methods, the design parameters for the various downhill and uphill

Table 4.2 Comparison of values of D, H_I and δ_H for the design of a horizontal sprinkler lateral (Example 1) and percentage differences between the design cases. Case (a): design for PVC pipe according to the proposed procedure and case (b): design for aluminum pipe according to the R-SBS method (Hathoot et al. 1994)

Design parameters	Proposed design procedure			R-SBS method (Hathoot et al. 1994)			Differences between the design cases (a) and (b)		
	Case (a): design for PVC pipe $F_S = 1.153$ $F_S = 1.203$ $F_S = 1.279$			Case (b): design for Aluminum pipe ($\varepsilon = 0.1$ mm)					
	$U_C = 76.7\%$ (1)	$U_C = 95.8\%$ (2)	$U_C = 98.9\%$ (3)	$U_C = 76.7\%$ (1)	$U_C = 95.8\%$ (2)	$U_C = 98.9\%$ (3)	(1)	(2)	(3)
D (mm)	51.3	75.3	101.6	50.8	76.2	101.6	–	–	–
H_I (m)	70.5	38.3	33.2	95.8	40.9	33.8	Increasing in H_I (%)		
							35.9	6.8	1.8
δ_H (%)	178.5	30.3	7.6	221.5	35.1	8.6	Increasing in δ_H (%)		
							24.1	15.6	12.5

line slope ranges are compared with those of the horizontal line slope case and, percentage differences in the pipe diameter and in the total energy loss for each slope combination are presented, respectively.

On the other hand, variation of the pipe diameter, D, and the total energy loss, $H_{T(L)}$, versus the presented slope combinations, are also shown in Fig. 4.1a and b, respectively. Figure 4.1a illustrates that, the pipe diameter decreases with increasing the down slope ranges whereas it increases with increasing the up slope ranges, with respect to the horizontal pipe case. From Table 4.3 for both the methods, the great percentage decrease in D is found to be 11% and 10%, respectively, for the highest down slope range, $S_0 = --0.05$, whereas the large increase in D is found to be 31.7% and 31.9%, respectively, for the up slope range, $S_0 = 0.03$. Table 4.3 shows that, the results of the design parameters are in a good agreement with a little deviation according to both the methods, for different slope combinations.

Figure 4.1b also illustrates that, the total energy loss increases due to pipe size decreasing as the down slope increases whereas it decreases due to pipe size increasing with increasing the up slope. Table 4.3 presents for both the methods, the large percentage increase in $H_{T(L)}$ that occurred is 69.2% and 71.5%, respectively, for the highest down slope range, $S_0 = -0.05$, whereas the large decrease in $H_{T(L)}$ is found to be 72.7% and 75.3%, respectively, for the up slope range, $S_0 = 0.03$.

As can be seen in Table 4.3, two realistic solutions of diameters exist for the down slopes, $S_0 = -0.04$ and $S_0 = -0.05$ and above; and both the diameter values (D = 29.14 mm and 16.61 mm) yields the same desired uniformity level, $U_C = 98.1\%$. On the other hand, for the up slope, $S_0 = 0.04$ and above, both the diameters are imaginary, indicating realistic solutions don't exist for this up slope range.

Figure 4.2a and b show the downstream end pressure head and operating inlet pressure head profiles, H_N and H_I versus the given slope combinations. Figure 4.2a illustrates, the downstream pressure head increases as the down slope increases, whereas it decreases as the up slope increases, with respect to the horizontal pipe case. Figure 4.2b also illustrates, the operating inlet pressure head decreases as the down slope increases, whereas it increases as the up slope increases when $S_0 \leq 0.02$, then it starts to decrease from the above of this up slope range.

Example 3: Design of Microirrigation Lateral Length

It is required to design the full length, L, of a horizontal PE microirrigation lateral, as well as the operating pressure head at the pipe inlet, H_I, and the amount of total energy losses, $H_{T(L)}$, considering the both design cases **(a) without local energy loss** ($F_S = 1.0$) and, **(b) including local energy loss** ($F_S = 1.15$) for $U_C = 95.2\%$ using the following data very similar to the one presented by Hathoot et al. (2000): The local energy loss of a labyrinth on-line emitter model is characterized by $l_e = 0.15$ m [17]. The internal diameter of the lateral, D = 13 mm; the spacing between successive emitters, s = 1.0 m with the inlet spacing ratio, $\mu = s_I/s = 1.0$; the emitter discharge exponent, x = 0.53; the average emitter

Table 4.3 Results for the design parameters D, H_I, H_N, $H_{f(L)}$ and $H_{T(L)}$ for the design of microirrigation lateral (Design Example 2) obtained with the proposed design procedure and those obtained from the R-SBS method (Hathoot et al. 2000) for different slope combinations ($S_0 = 0.0, \mp 0.005, \mp 0.01, \mp 0.02, \mp 0.03, \mp 0.04$, and ∓ 0.05), and percentage differences of D and $H_{T(L)}$ with respect to the design case of a horizontal pipe. *Second larger diameter obtained from Eq. (4.14a)

Proposed design procedure								R-SBS method					
Slope (%)	D (mm)	H_I(m)	H_N(m)	$H_{f(L)}$(m)	$H_{T(L)}$ (m)	Decrease (%)	Increase (%)	D (m)	H_I (m)	H_N (m)	$H_{T(L)}$ (m)	Decrease (%)	Increase (%)
						D (mm)	$H_{T(L)}$ (m)					D (mm)	$H_{T(L)}$ (m)
Leveled-0	13.40	11.65	8.16	2.79	3.52	–	–	13.00	11.50	8.39	3.16	–	–
Down-0.5	13.13	11.59	8.27	3.07	3.87	2.0	9.9	12.70	11.47	8.48	3.54	2.3	12.0
Down-1	12.90	11.54	8.39	3.32	4.19	3.7	19.1	12.55	11.31	8.60	3.77	3.5	19.3
Down-2	12.53	11.39	8.65	3.80	4.79	6.5	36.1	12.33	10.90	8.83	4.14	5.2	31.0
Down-3	12.25	11.19	8.94	4.21	5.31	8.6	50.8	12.05	10.64	9.07	4.65	7.3	47.2
Down-4	12.06 (29.14)*	10.91	9.29	4.51	5.68	10.0	61.5	11.86	10.38	9.43	5.04	8.8	59.5
Down-5	11.93 (16.61)*	10.56	9.69	4.72	5.95	11.0	69.2	11.70	10.10	9.79	5.42	10.0	71.5
						Increase (%)	Decrease (%)					Increase (%)	Decrease (%)
						D (mm)	$H_{T(L)}$ (m)					D (mm)	$H_{T(L)}$ (m)
Up–0.5	13.73	11.67	8.07	2.48	3.12	2.5	11.3	13.36	11.53	8.31	2.77	2.8	12.3
Up–1	14.13	11.70	7.97	2.18	2.75	5.5	21.9	13.68	11.66	8.23	2.45	5.2	22.5
Up–2	15.28	11.72	7.81	1.51	1.90	14.0	45.9	14.92	11.70	8.11	1.59	14.8	49.7

(continued)

Table 4.3 (continued)

Proposed design procedure								R-SBS method					
Slope (%)	D (mm)	H_I(m)	H_N(m)	$H_{f(L)}$(m)	$H_{T(L)}$ (m)	Decrease (%)	Increase (%)	D (m)	H_I (m)	H_N (m)	$H_{T(L)}$ (m)	Decrease (%)	Increase (%)
						D (mm)	$H_{T(L)}$ (m)					D (mm)	$H_{T(L)}$ (m)
Up−3	17.66	11.67	7.69	0.76	0.96	31.7	72.7	17.15	11.68	7.88	0.78	31.9	75.3
Up−4	a	–	–	–	–	–	–	a	–	–	–	–	–
Up−5	a	–	–	–	–	–	–	a	–	–	–	–	–

[a]No real root is found from Eq. (4.14a)

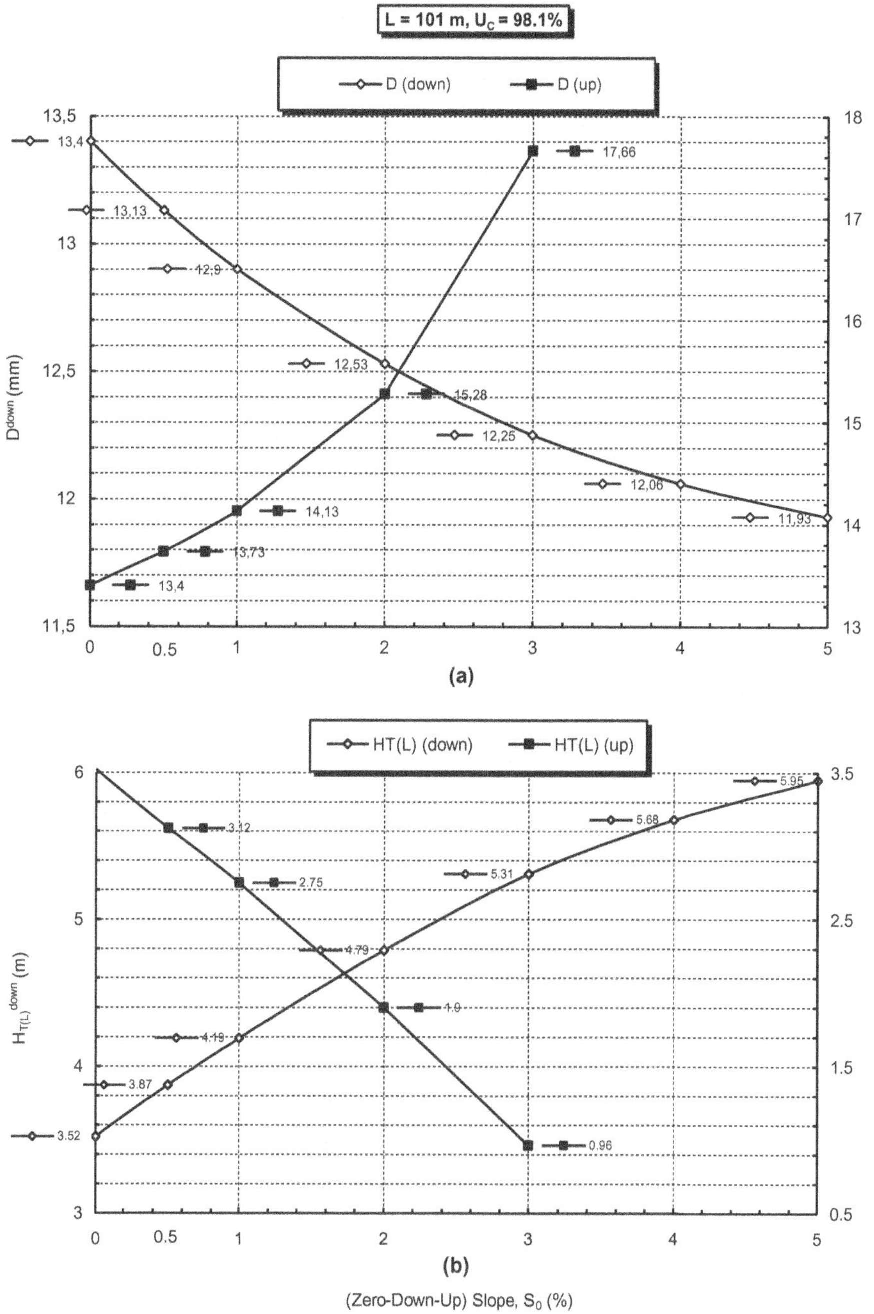

Fig. 4.1 Design chart for the design Example 2: for L = 101 m, $U_C = 98.1\%$, and for various slope combinations ($S_0 = 0.0, \mp 0.005, \mp 0.01, \mp 0.02, \mp 0.03, \mp 0.04$, and ∓ 0.05); **a**: variation of pipe diameter (D^{down}, D^{up}), and **b**: variation of total energy losses ($H_{T(L)}{}^{down}$, $H_{T(L)}{}^{up}$), respectively, according to the proposed design procedure

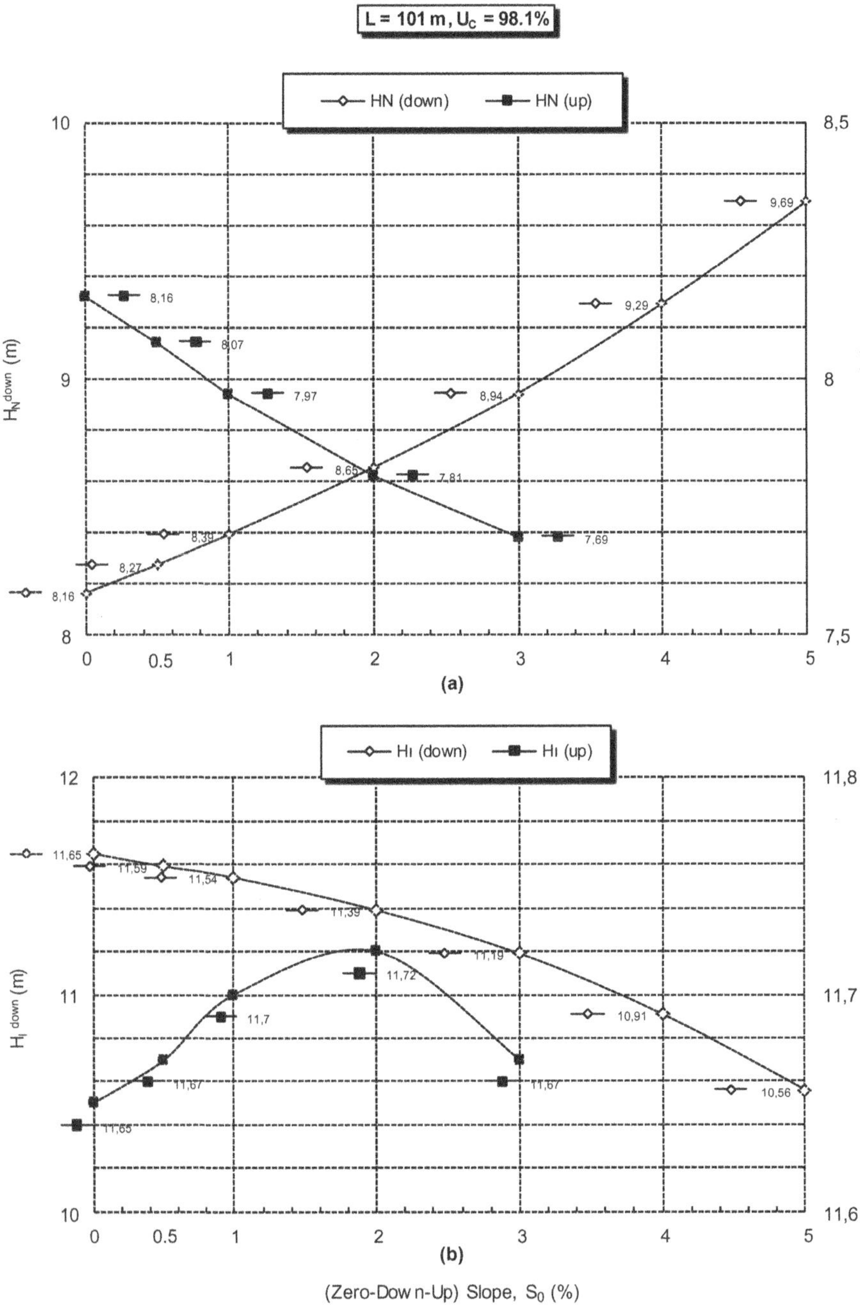

Fig. 4.2 Design chart for the design Example 2: for L = 101 m, $U_C = 98.1\%$, and for various slope combinations ($S_0 = 0.0, \mp 0.005, \mp 0.01, \mp 0.02, \mp 0.03, \mp 0.04$, and ∓ 0.05); **a**: variation of downstream end pressure head ($H_N{}^{down}$, $H_N{}^{up}$), and **b**: variation of operating inlet pressure head ($H_I{}^{down}$, $H_I{}^{up}$), respectively, according to the proposed design procedure

discharge $\overline{q} = 8.0$ lph (2.222×10^{-6} m^{3}s^{-1}) and corresponding pressure head, $\bar{H} = 9.6$ m, respectively.

According to the presented design procedure, the design case (**a**) gives the results: L = 69.2 m, $H_I = 12.35$ m, and $H_{T(L)} = 3.76$ m, whereas the design case (**b**) yields L = 63.6 m, $H_I = 11.84$ m, and $H_{T(L)} = 3.51$ m, respectively. It is obvious that, considering local energy losses has the effect of decreasing the pipe length 8.1%, decreasing the operating inlet pressure head 4.1%, as well as decreasing the total energy loss due to pipe length by 6.7%, respectively. For this example, the R-SBS numerical method [13] gives results similar for the pipe length; for the case of neglecting emitter losses, L = 71.0 m and for the case of including emitter losses for small, medium and large sizes of emitter barb, L = 68.0 m, 67.0 m and 64.0 m, respectively. It means according to the R-SBS design method, taking into account the emitter local losses decreases the pipe length with respect to the case of neglecting emitter local losses, 4.2%, 5.6% and 10%, respectively, as earlier reported in the mentioned literature [13].

For the extension of the above design problem, it is solved with the same data for various slope combinations (S_0 = 0.0, ∓0.005, ∓0.01, ∓0.02, ∓0.03, ∓0.04, and ∓ 0.05), as well. For the sake of comparison, the complete results for the design parameters obtained from the proposed method and the R-SBS method with the medium size of emitter barb are presented in Table 4.4. As seen from this table for both the methods, the design parameters for the various downhill and uphill line slope ranges are compared with those of the horizontal line slope case and, percentage differences in the pipe length and in the total energy loss for each slope combination are presented, respectively.

In addition, variation of the pipe length, L, and the total energy loss, $H_{T(L)}$, versus the given slope combinations, are indicated in Fig. 4.3a and b, respectively. Figure 4.3a illustrates that, the pipe length increases with increasing the down slope range whereas it decreases with increasing the up slope range with respect to the horizontal pipe case. From Table 4.4 for both the methods, the greatest percentage increase in L of 25.8% and 23.1%, respectively, for the highest down slope range, $S_0 = -0.05$; whereas the greatest decrease in L of 26.6% and 25.7%, respectively, for the highest up slope range, $S_0 = 0.05$, are obtained.

Figure 4.3b also illustrates, the total energy loss increases due to pipe length increasing as the down slope increases whereas it decreases due to pipe length decreasing with increasing the up slope. Table 4.4 indicates for both the methods, the largest percentage increase in $H_{T(L)}$ of 84.7% and 84.7%, respectively, for the highest down slope range, $S_0 = -0.05$; whereas the greatest percentage decrease in $H_{T(L)}$ of 57.2% and 60.5%, respectively, for the highest up slope range, $S_0 = 0.05$, are found.

Figure 4.4a and b show the downstream end pressure head and operating inlet pressure head profiles, H_N and H_I versus the given slope combinations. Figure 4.4a illustrates that, the downstream pressure head increases as the down slope increases, whereas it decreases as the up slope increases, with respect to the horizontal pipe case. As concluded

Table 4.4 Results for the design parameters L, H_I, H_N, $H_{f(L)}$ and $H_{T(L)}$ for the design of microirrigation lateral (Design Example 3) obtained with the proposed design procedure and those obtained from the R-SBS method (Hathoot et al. 2000) for different slope combinations

	Proposed design procedure							R-SBS method					
Slope[a] (%)	L (m)	H_I (m)	H_N (m)	$H_{f(L)}$ (m)	$H_{T(L)}$ (m)	Increase (%)		L (m)	H_I (m)	H_N (m)	$H_{T(L)}$ (m)	Increase (%)	
						L (m)	$H_{T(L)}$ (m)					L (m)	$H_{T(L)}$ (m)
Leveled-0	63.6	11.84	8.39	3.05	3.51	–	–	67.0	11.80	8.52	3.34	–	–
Down-0.5	66.3	11.87	8.44	3.32	3.82	4.2	8.9	69.4	11.82	8.59	3.64	3.6	8.9
Down-1	68.1	11.91	8.50	3.61	4.15	7.1	18.2	70.8	11.84	8.67	3.96	5.7	18.6
Down-2	71.3	11.87	8.69	4.06	4.67	12.1	33.1	74.6	11.83	8.76	4.65	11.3	39.2
Down-3	74.5	11.94	8.84	4.72	5.43	17.1	54.7	77.1	11.63	8.96	5.06	15.1	51.5
Down-4	77.4	11.79	9.11	5.08	5.84	21.7	66.3	79.4	11.40	9.14	5.51	18.5	64.9
Down-5	80.0	11.74	9.34	5.64	6.49	25.8	84.7	82.5	11.32	9.35	6.17	23.1	84.7
						Decrease (%)						Decrease (%)	
						L (m)	$H_{T(L)}$ (m)					L (m)	$H_{T(L)}$ (m)
Up-0.5	62.7	11.90	8.28	2.92	3.36	1.4	4.2	65.8	11.95	8.51	3.17	1.8	5.1
Up -1	60.9	11.87	8.23	2.68	3.08	4.2	12.4	63.9	11.90	8.41	2.91	4.6	12.9
Up -2	57.3	11.80	8.15	2.22	2.55	9.9	27.3	60.4	11.89	8.35	2.39	9.9	28.4
Up -3	53.7	11.81	8.03	1.91	2.20	15.6	37.3	57.2	12.30	8.63	2.01	14.6	39.8
Up -4	50.1	11.74	7.99	1.55	1.78	21.2	49.3	53.3	12.25	8.56	1.61	20.4	51.8
Up -5	46.7	11.73	7.91	1.31	1.50	26.6	57.2	49.8	12.28	8.50	1.32	25.7	60.5

[a]($S_0 = 0.0, \mp 0.005, \mp 0.01, \mp 0.02, \mp 0.03, \mp 0.04$, and ∓ 0.05), and percentage differences of L and $H_{T(L)}$ with respect to the design case of a horizontal pipe

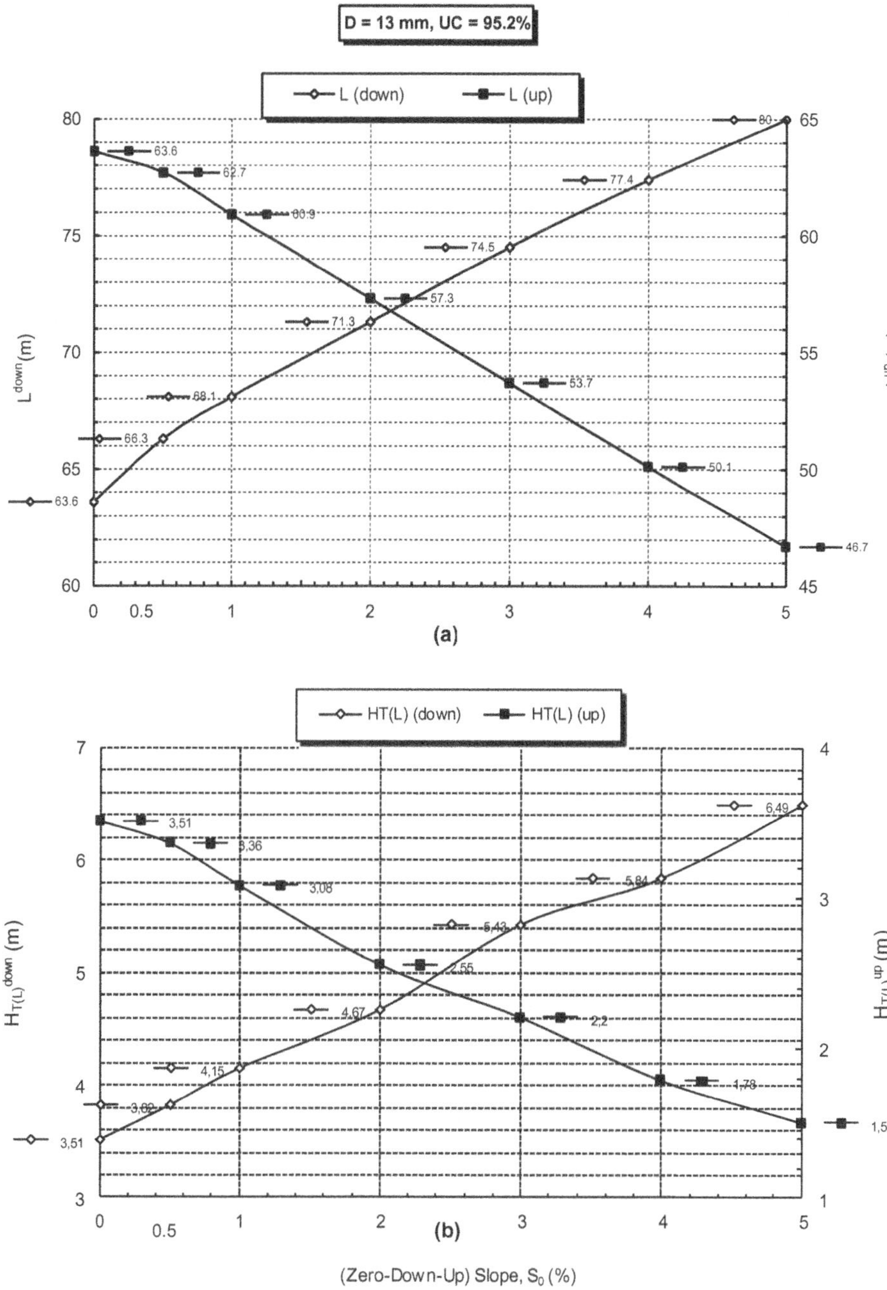

Fig. 4.3 Design chart for the design Example 3: for D = 13 mm, U_C = 95.2%, and for various slope combinations ($S_0 = 0.0, \mp 0.005, \mp 0.01, \mp 0.02, \mp 0.03, \mp 0.04$, and ∓ 0.05); **a**: variation of pipe length (L^{down}, L^{up}), and **b**: variation of total energy losses ($H_{T(L)}{}^{down}$, $H_{T(L)}{}^{up}$), respectively, according to the proposed design procedure

in Fig. 4.4b or in Table 4.4, no tendency is observed for the operating inlet pressure head profile and its values fluctuates depending on different downhill and uphill slope range.

Concluding Remarks

In this study, an analytical procedure for hydraulic analysis and solving direct design problem of a single multiple outlets pipeline, is presented. The proposed equations, taking into consideration the influence of local energy loss, can be applied for various types of outlet, different flow regimes and different uniform line slope ranges. In the procedure, for any desired uniformity level and given design slope range with remaining known parameters, the pipe diameter and the pipe length can then be directly designed. For any desired uniformity level, the procedure also provides an opportunity to evaluate the influence of local energy loss, as well as the influence of different uniform line slopes on the pipe geometric characteristics (pipe size and length), and on the corresponding hydraulic variables (operating inlet pressure head, downstream end pressure head and total energy loss). The present technique can be applied to laterals and manifolds in both sprinkle and trickle systems practically under any situations, and showing in this way its great flexibility. It may be concluded that, the performance of the presented technique is found to be sufficiently accurate in all cases examined even for uphill slope condition in comparison with a revised step-by-step (R-SBS) method. The procedure can be applied easily for practical purposes without an excessive calculation effort, and preferred instead for the numerical techniques which need a very long execution time and complex computer operation.

Based on the present research, the following conclusions are underlined:

1. Examination of the results for sprinkler lateral design shows that, the influence of coupler energy loss on the pipe size, operating pressure head and total energy loss, can become significant and neglecting the coupler local loss may lead to designs in error. For trickle lateral design, considering the effect of emitter connection losses may lead to increasing the pipe diameter 30.1%, to decreasing the operating inlet pressure head 30.4%, as well as to decreasing the total energy loss due to pipe size increasing 63.9%, respectively. On the other hand, considering the effect of emitter connection losses may lead to decreasing the pipe length 8.1%, to decreasing the operating inlet pressure head 4.1%, as well as to decreasing the total energy loss due to pipe length decreasing 6.7%, respectively.
2. Examination of the results for trickle lateral design indicates that, as the down slope increases the pipe diameter decreases and the total energy loss increases due to pipe size decreasing, whereas the pipe diameter increases and the total energy loss decreases due to pipe size increasing with increasing the up slope, in comparison with the design case of horizontal pipe. On the other hand, as the down slope increases the pipe length

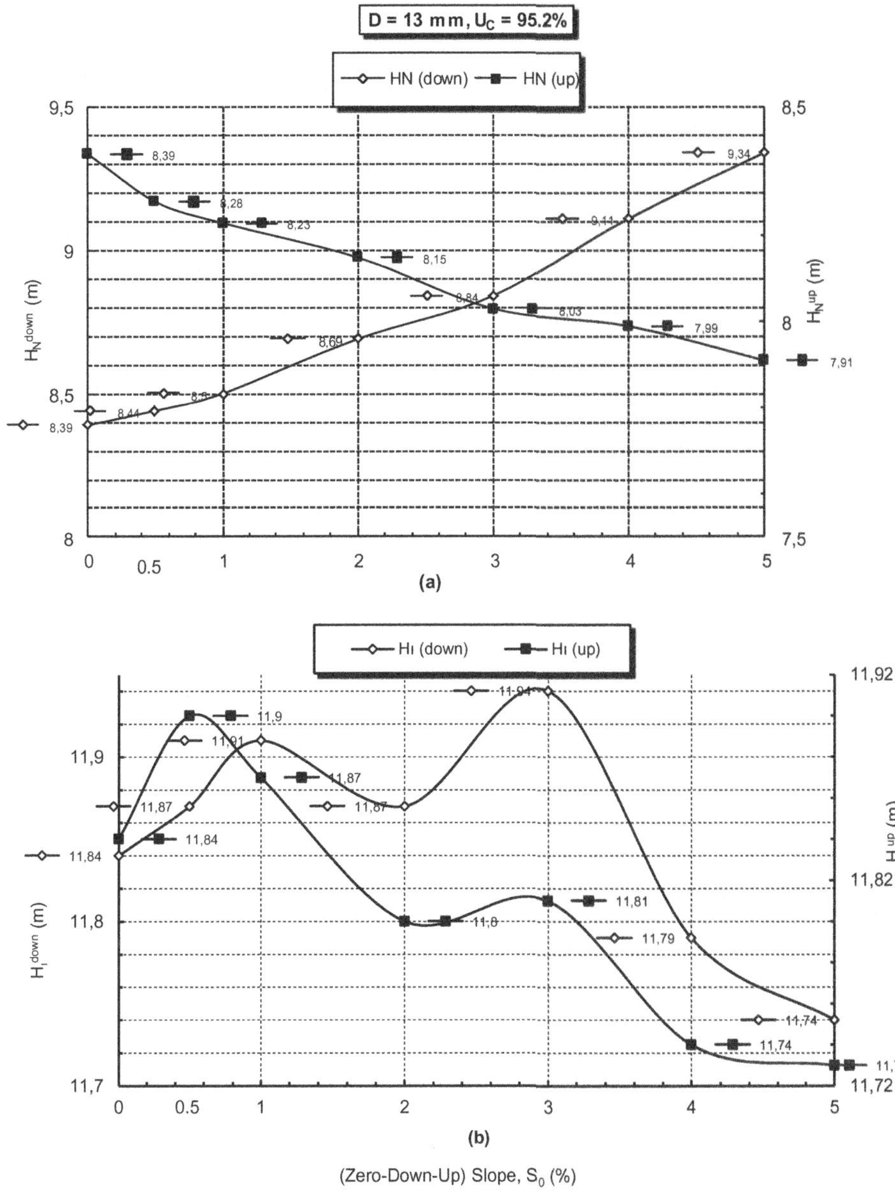

Fig. 4.4 Design chart for the design Example 3: for D = 13 mm, $U_C = 95.2\%$, and for various slope combinations ($S_0 = 0.0, \mp 0.005, \mp 0.01, \mp 0.02, \mp 0.03, \mp 0.04$, and ∓ 0.05); **a**: variation of downstream end pressure head ($H_N{}^{down}$, $H_N{}^{up}$), and **b**: variation of operating inlet pressure head ($H_I{}^{down}$, $H_I{}^{up}$), respectively, according to the proposed design procedure

increases and the total energy loss increases due to pipe length increasing whereas the pipe length decreases and the total energy loss decreases due to pipe length decreasing with increasing the up slope, in comparison with respect to the design case of horizontal pipe.

Acknowledgements The senior author wants to express his acknowledges to Prof. Vijay. P. Singh at Texas A&M University, Emeritus Prof. Necati Ağıralioğlu from International Antalya Science University, and Distinguished Emeritus Professor Zekai Şen from Istanbul Medipol University, for their highly encouragement on preparing this book, many scientific activities, projects and so on. The Scientific and Technological Research Council of Turkey (TUBITAK) is acknowledged for supporting by the fellowships and grant program during the studies in the Texas A&M University/BAEN. We also thank to publishers, potential readers and reviewers for providing constructive reviews and criticism on the content of the book, in advance.

Conflict of Interest The author/editor has no conflicts of interest, financial or otherwise, to declare.

Appendices

Appendix 1: Determination of Energy-Gradient Line (EGL)

Let H(l) be the pressure head at a given length l from the inlet, H_I be the pressure head at the pipeline inlet, and S_0 be the uniform slope of the lateral line.

Assuming the average pressure head for the entire pipeline is located in the middle of the length of the pipeline, and applying conservation of energy principle between the pipe segments $l = 0$ and $l = l$, then pressure head distribution along the pipeline is expressed by

$$H(l) = H_I + H_{VI}\left[1 - (1 - i)^2\right] - F_S H_{f(L)}\left[1 - (1 - i)^{m+1}\right] - pi\Delta S \tag{4.16}$$

The average pressure head along the pipeline, can then be written [5, 70–72]:

$$\overline{H} = H_I - (\overline{H}_f + p\overline{\Delta S}) \tag{4.17}$$

in which $\overline{H}_f$ = average head loss due to friction along the pipeline (m); and $\overline{\Delta S} = \Delta S/2$ = average head loss or gain due to elevation change for the entire pipeline (m).

The average head loss due to friction along the pipeline, $\overline{H}_f$ is expressed [46]

$$\overline{H}_f = \frac{1}{L}\int_{l=0}^{l=L} dS = \frac{1}{L}\int_{l=0}^{l=L} H_f(l)dl = \left(\frac{m+1}{m+2}\right)H_{f(L)} \tag{4.18}$$

Substituting the expressions for $\overline{\Delta S}$, and $\overline{H}_f$ presented above into Eq. (4.17) then, subtracting from Eq. (4.16) finally, the result becomes to Eq. (4.7) in the text.

Appendix 2: Determination of Nonuniform Outflow Exponent (B)

To alternate the traditional relationship for the linear decreasing flow rate through the pipeline given by Eq. (4.1), to describe the variable flow rate along the pipeline Q(l), simple but sufficiently accurate exponential function can be used [20, 21, 51, 52]

$$Q(l) = Q_I(1 - i)^{\beta} \tag{4.19}$$

If Eq. (4.19) is evaluated for the upper half segment of the pipeline ($i = 0.5$), then arranged with the expression for the flow rate entering over the upper half segment of the pipeline, $Q_{0.5} = (kN/2)\overline{H}_{0.5}^{x}$ (k = coefficient of the outlet discharge) finally, the β can be evaluated from:

$$\beta = 1 - 3.322x \log\left(\frac{\overline{H}_{0.5}}{\overline{H}}\right) \tag{4.20}$$

where $\overline{H}_{0.5}$ = average pressure head along the upper half segment of the pipeline.

Assuming the $\overline{H}_{0.5}$ is located in the middle of the length of the upper half segment of the pipeline, and applying conservation of energy equation between the pipe segments l = L/4 and l = L, then the $\overline{H}_{0.5}$ can be expressed [60]:

$$\overline{H}_{0.5} = H_N + H_{f(L)} + (p\overline{\Delta S}_{0.5} - \overline{H}_{f0.5}) \tag{4.21}$$

where H_N = downstream pressure head at the first outlet from the closed end of the pipeline (m); $\overline{H}_{f0.5}$ = average head loss due to friction along the upper half segment of the pipeline (m); and $\overline{\Delta S}_{0.5}$ = average head loss or gain due to elevation change at the end of the upper-quarter segment of the pipeline (m), which is written by this form:

$$\overline{\Delta S}_{0.5} = \frac{3}{4}\Delta S \tag{4.22}$$

The H_N can be obtained for $i = 1$ from the pressure head profile given by Eq. (4.7) with taking into account the flow rate exponent, $m_\beta = 1.0 \times m$:

$$H_N = \overline{H} + H_{VI} + H_{f(L)}\left[\frac{m+1}{m+2} - F_S\right] - \frac{p\Delta S}{2} \tag{4.23}$$

Average head loss due to friction between the pipe segments $l = 0$ and l = L/2 for the upper half segment of the pipeline, $\overline{H}_{f0.5}$, is expressed by [56]:

$$\bar{H}_{f0.5} = \frac{2}{L}\int_{l=0}^{l=L/2} dS = \frac{2}{L}\int_{l=0}^{l=L/2} H_f(l)dl = \left(\frac{m + 0.5^{m+1}}{m+2}\right)H_{f(L)} \tag{4.24}$$

Substituting Eqs. (4.22), (4.23) and (4.24) into Eq. (4.21), and dividing by $\bar{H}$ then, substituting into Eq. (4.20) finally, the result transforms to Eq. (4.8) in the text.

It should be noted that, a similar form of Eq. (4.8) was presented in the literature [21]. However, Eq. (4.29) is essentially based on the average outlet flow approximation, which is earlier discussed in the recent papers [26–28]; whereas the average friction drop approach [2–5, 22, 34, 45, 48] is basically used along the background derivative procedure of the proposed analytical development (*Appendices I and II*).

References

1. A.I. Al-Amoud, Significance of energy losses due to emitter connections in trickle irrigation lines. J. Agric. Eng. Res. **60**(1), 1–5 (1995)
2. A.A. Anwar, Factor G for pipelines with equally spaced multiple outlets and outflow. J. Irrig. Drain. Eng. **125**(1), 34–39 (1999a)
3. A.A. Anwar, Adjusted factor G_a for pipelines with multiple outlets and outflow. J. Irrig. Drain. Eng. **125**(6), 355–359 (1999b)
4. A.A. Anwar, Inlet pressure for horizontal tapered laterals. J. Irrig. Drain. Eng. **126**(1), 57–63 (2000)
5. H. Anyoji, I.P. Wu, Statistical approach for drip lateral design. Trans. ASAE **30**(1), 187–192 (1987)
6. I.P. Wu, Energy gradient line approach for direct hydraulic calculation in drip irrigation design. Irrig. Sci. **13**, 21–29 (1992)
7. I.P. Wu, An assessment of hydraulic design of microirrigation systems. Agric. Water Manage. **32**, 275–284 (1997)
8. I.P. Wu, H.M. Gitlin, Drip irrigation design based on uniformity. Trans. ASAE **17**(3), 429–432 (1974)
9. I.P. Wu, H.M. Gitlin, Energy gradient line for drip irrigation laterals. J. Irrig. Drain. Eng. **101**(4), 323–326 (1975)
10. I.P. Wu, R. Yue, Drip lateral design using energy gradient line approach. Trans. ASAE **36**(2), 389–394 (1993)
11. H.M. Hathoot, A.I. Al-Amoud, F.S. Mohammad, Analysis and design of trickle irrigation laterals. J. Irrig. Drain. Eng. **119**(5), 756–767 (1993)
12. H.M. Hathoot, A.I. Al-Amoud, F.S. Mohammad, Analysis and design of sprinkler irrigation laterals. J. Irrig. Drain. Eng. **120**(3), 534–549 (1994)
13. H.M. Hathoot, A.I. Al-Amoud, A.S. Al-Mesned, Design of trickle irrigation laterals considering emitter losses. ICID J **49**(2), 1–14 (2000)
14. G. Yıldırım, Hydraulic analysis and direct design of multiple outlets pipelines laid on flat and sloping lands. J. Irrig. Drain. Eng. ASCE. **132**(6), 537–552 (2006)
15. G. Yıldırım, An Assessment for hydraulically efficient design of uniformly sloping submain, lines. Irrig. Drain. **64**(2), 157–168 (2015)

16. G. Yıldırım, Total energy loss assessment for trickle lateral lines equipped with integrated in-line and on-line emitters. Irrig. Sci. **28**(4), 341–352 (2010)
17. Y. Kang, S. Nishiyama, Analysis and design of microirrigation laterals. J. Irrig. Drain. Eng. **122**(2), 75–82 (1996a)
18. Y. Kang, S. Nishiyama, Design of microirrigation submain unit. J. Irrig. Drain. Eng. **122**(2), 83–89 (1996b)
19. Y. Kang, B. Yuan, S. Nishiyama, Design of microirrigation laterals at minimum cost. Irrig. Sci. **28**, 125–133 (1999)
20. J.D. Valiantzas, Analytical approach for direct drip lateral hydraulic calculation. J. Irrig. Drain. Eng. **124**(6), 300–305 (1998)
21. J.D. Valiantzas, Continuous outflow variation along irrigation laterals: effect of the number of outlets. J. Irrig. Drain. Eng. **128**(1), 34–42 (2002)
22. G. Yıldırım, N. Ağıralioğlu, Discussion of 'continuous outflow variation along irrigation laterals: effect of the number of outlets' by J. D. Valiantzas. J. Irrig. Drain. Eng. **129**(5), 382–386 (2003)
23. G. Yıldırım, N. Ağıralioğlu, Linear solution for hydraulic analysis of tapered microirrigation laterals, J. Irrig. and Drain. Eng. **130**(1), 78–87 (2004a)
24. G. Yıldırım, N. Ağıralioğlu, Comparative analysis of hydraulic calculation methods in design of microirrigation laterals. J. Irrig. Drain. Eng. **130**(3), 201–217 (2004b)
25. V. Ravikumar, C.R. Ranganathan, S. Santhana Bosu, Analytical equation for variation of discharge in drip irrigation laterals. J. Irrig. Drain. Eng. **129**(4), 295–298 (2003)
26. G. Yıldırım, N. Ağıralioğlu, Discussion of 'inlet pressure, energy cost, and economic design of tapered irrigation submains' by J. D. Valiantzas. J. Irrig. Drain. Eng. **131**(2), 220–224 (2005a)
27. G. Yıldırım, N. Ağıralioğlu, Closure to 'linear solution for hydraulic analysis of tapered microirrigation laterals' by G. Yıldırım and N. Ağıralioğlu. J. Irrig. Drain. Eng. **131**(5), 490–491 (2005b)
28. G. Yıldırım, N. Ağıralioğlu, Closure to 'Comparative analysis of hydraulic calculation methods in design of microirrigation laterals' by G. Yıldırım and N. Ağıralioğlu. J. Irrig. Drain. Eng. **132**(1), 85–90 (2006)
29. V. Bagarello, V. Ferro, G. Provenzano, D. Pumo, Evaluating pressure losses in drip irrigation lines. J. Irrig. Drain. Eng. **123**(1), 1–7 (1997)
30. L. Juana, L. Rodriguez-Sinobas, A. Losada, Determining minor head losses in drip irrigation laterals. I: methodology. J. Irrig. Drain. Eng. **128**(6): 376–384 (2002a).
31. L. Juana, L. Rodriguez-Sinobas, A. Losada, Determining minor head losses in drip irrigation laterals. II: experimental study and validation. J. Irrig. Drain. Eng. **128**(6): 385–396 (2002b).
32. L. Juana, A. Losada, A.L. Rodrigoez-Sinobas, R. Sanchez, Analytical relationships for designing rectangular drip irrigation units. J. Irrig. Drain. Eng. **130**(1), 47–59 (2004)
33. G. Provenzano, D. Pumo, Experimental analysis of local pressure losses for microirrigation laterals. J. Irrig. Drain. Eng. **130**(4), 318–324 (2004)
34. G. Provenzano, D. Pumo, P. Di Dio, Simplified procedure to evaluate head losses in drip irrigation laterals. J. Irrig. Drain. Eng. **131**(6), 525–532 (2005)
35. G. Provenzano, G. Palau-Salvador, Discussion of 'Comparative analysis of hydraulic calculation methods in design of microirrigation laterals' by G. Yıldırım and N. Ağıralioğlu. J. Irrig. Drain. Eng. **132**(1), 82–85 (2005)
36. G. Yıldırım, Discussion of 'Experimental analysis of local pressure losses for microirrigation laterals' by G. Provenzano and D. Pumo. J. Irrig. Drain. Eng. **132**(2): 189–193 (2006b)
37. G. Yıldırım, Simplified procedure for hydraulic design of small-diameter plastic pipes. Irrig. Drain. (ICID J) **58**(3), 209–233 (2009)

38. G. Yıldırım, An assessment of hydraulic design of trickle laterals considering effect of minor losses. Irrig Drain (ICID J) **56**(4), 399–421 (2007)
39. G. Yıldırım, Discussion of "Statistical analysis of non-pressure-compensating and pressure-compensating drip emitters" by H. Perea, J. Enciso-Medina, V. P. Singh, D. P. Dutta, and B. J. Lesikar. J. Irrig. Drain. Eng., ASCE (2016). https://doi.org/10.1061/(ASCE)IR.1943-4774.0000644
40. G. Yıldırım, Analytical relationships for designing multiple outlets pipelines. J. Irrig. Drain. Eng. (2006a)
41. J.E. Christiansen, *Irrigation by sprinkling.* (California Agricultural Experimental Station Bulletin 670, University of California, Davis, California, 1942)
42. S.W. Churchill, Friction factor equation spans all fluid-flow regimes. Chem. Eng. **84**(24), 91–92 (1977)
43. R.H. Cuenca, *Irrigation systems design: an engineering approach* (Prentice Hall, New Jersey, 1989)
44. E.J. Finnemore, J.B. Franzini, *Fluid mechanics with engineering applications*, 10th edn. (McGraw-Hill, 2002), pp. 745–754
45. S.K. Jain, K.K. Singh, R.P. Singh, Microirrigation lateral design using lateral discharge equation. J. Irrig. Drain. Eng. **128**(2), 125–128 (2002)
46. M.E. Jensen, Design and operation of farm irrigation systems. ASAE Monograph. St. Joseph, Mich, 1981
47. J. Keller, D. Karmelli, Trickle irrigation design parameters. Trans. ASAE **17**(4), 678–684 (1974)
48. P.S. Mahar, R.P. Singh, Computing inlet pressure head of multioutlet pipeline. J. Irrig. Drain. Eng. **129**(6), 464–467 (2003)
49. D.S. Miller, *Internal flow systems.* (*BHRA* Publications, Cranfield, U. K., 1990), pp. 189–203
50. A.C. Sarsfield, *Irrigation technical manual: engineering data.* (Irrigation Technical Services, Lafayette, Calif, 1984)
51. E.J. Scaloppi, R.G. Allen, Hydraulics of irrigation laterals: comparative analysis. J. Irrig. Drain. Eng. **119**(1), 91–115 (1993)
52. P. Swamee, A. Jain, Explicit equations for pipe flow problems. J. Hydraulic Div. 657–664 (1976)
53. P. Vallesquino, P.L. Luque-Escamilla, New algorithm for hydraulic calculation in irrigation laterals. J. Irrig. Drain. Eng. **127**(4), 254–260 (2001)
54. P. Vallesquino P.L. Luque-Escamilla, Equivalent friction factor method for hydraulic calculation in irrigation laterals. J. Irrig. Drain. Eng. **128**(5), 278–286 (2002)
55. R.D. von Bernuth, Simple and accurate friction loss equation for plastic pipe. J. Irrig. Drain. Eng. **116**(2), 294–329 (1990)
56. R.D. von Bernuth, T. Wilson, Friction factors for small diameter plastic pipes. J. Hydraulic Eng. **115**(2), 183–192 (1989)
57. A.W. Warrick, M. Yitayew, An analytical solution for flow in a manifold. Adv. Water Resources **10**, 58–63 (1987)
58. A.W. Warrick, M. Yitayew, Trickle lateral hydraulics. I: analytical solution. J. Irrig. Drain. Eng. **114**(2), 281–288 (1988)
59. G.Z. Watters, J. Keller, Trickle irrigation tubing hydraulics, in *presented at the Summer Meeting of ASAE* (Utah State University, Logan, Utah, 1978), Paper No. 78–2015
60. M. Yitayew, Head loss in manifolds or trickle lateral: simplified approach. J. Irrig. Drain. Eng. **115**(4), 739–743 (1989)
61. M. Yitayew, A.W. Warrick, Trickle lateral hydraulics. II: design and examples. J. Irrig. Drain. Eng. **114**(2), 289–300 (1988)

62. G. Yıldırım, O. Kose, New software (Multi-flowCAD) development for sub-main line hydraulic design: computational algorithm, computer visualization and implementation. Arab. J. Sci. Eng. **41**, 1551–1560 (2016)
63. G.Yıldırım, Computer-based analysis of hydraulic design variables for uniformly sloping microirrigation system laterals. J. Irrig. Drain. Eng., ASCE. **141**(7), 04014079 (2015)
64. G. Yıldırım, V.P. Singh, A *MathCAD* procedure for commercial pipeline hydraulic design considering local energy losses. Adv. Eng. Softw. **41**(2), 489–496 (2010)
65. G. Yıldırım, Computer-based analysis of explicit approximations to the implicit colebrook-white equation in turbulent flow friction factor calculation. Adv. Eng. Softw. **40**(11), 1183–1190 (2009)
66. G. Yıldırım, M. Özger, Neuro-fuzzy approach in estimating hazen-williams friction coefficient for small-diameter polyethylene pipes. Adv. Eng. Softw. **40**(8), 593–599 (2009)
67. M. Özger, M.G. Yıldırım, Determining turbulent flow friction coefficient using adaptive neuro-fuzzy computing technique. Adv. Eng. Softw. **40**(4), 281–287 (2009)
68. G. Yıldırım, Determining operating inlet pressure head incorporating uniformity parameters for multioutlet plastic pipelines. J. Irrig. Drain. Eng., ASCE. **134**(3), 341–348 (2008)
69. G. Yıldırım, Analytical relationships for designing multiple outlets pipelines. J. Irrig. Drain. Eng. ASCE. **133**(2), 140–154 (2007)
70. I.P. Wu, Personal communication (2002)
71. G. Yıldırım, *Comparative analysis of hydraulic calculation methods in design oh micro-irrigation laterals and linear solution of hydraulic analysis of tapered laterals.* (Istanbul Technical University, Hydraulics Division, Istanbul, Turkey, 2007). PhD. Thesis
72. G. Yıldırım, *Computer-aided design of micro-irrigation laterals by using LatCAD* (Thesis, Istanbul Technical University, Hydraulics Division, Istanbul, Turkey, MSc, 2001)

5 Analysis of Total Energy and Pressure Losses Components Along Micro-irrigation Laterals Equipped with Integrated In-Line and On-Line Emitters: Backward Stepwise (B-SBS) Procedure

Introduction

In a trickle irrigation system, water is distributed through small dissipating devices called emitters installed on polyethylene pipes called laterals [1–12]. Lateral pipeline is a hydraulic structure whose design is limited by the inlet pressure head and water application uniformity that is affected by the total energy losses, the field topography as well as the emitter hydraulic characteristics [13–24]. The insertion of emitters along a trickle lateral modifies the flow streamlines, inducing local turbulence that results in additional local pressure losses, sometimes called minor losses, rather than the pipe friction losses [25, 26]. To accurately evaluate total energy losses in the laterals, these minor losses due to the protrusion of emitter connections in the pipe wall that must be added to the friction losses occurring in the pipe [14]. On-line emitters cause the contraction and subsequent enlargement of flow streamlines due to the protrusion of emitter bars into the flow [16, 23, 25]. The introduction of integrated in-line emitters determines the contraction of the flow paths at the upstream connection between the emitters and the lateral pipe, and the expansion of the flow paths immediately downstream from the emitters; because the emitters usually have a smaller diameter than the pipe, an additional minor friction losses must be considered [9–11]. Figure 5.1 presents the lateral-emitter configuration for (**a**): **integrated in-line** and (**b**): **on-line emitters.**

In the past, numerous researches have been done on the hydraulic analysis of trickle irrigation pipeline networks [26–40]. As a matter of fact, a significant amount of these researches [13, 27–34] do not taken into account the effect of local losses in the design procedure, although the importance of minor losses has been recently presented in the experimental analysis [3–11, 18, 20], and the alternative analytical and numerical approaches [5, 14, 23, 25, 30, 41].

G. Yıldırım, *Hydraulic Principles and Design Concepts for Submain Units with Multiple Outlet Pipelines*, Synthesis Lectures on Engineering, Science, and Technology,
https://doi.org/10.1007/978-3-031-32495-6_5

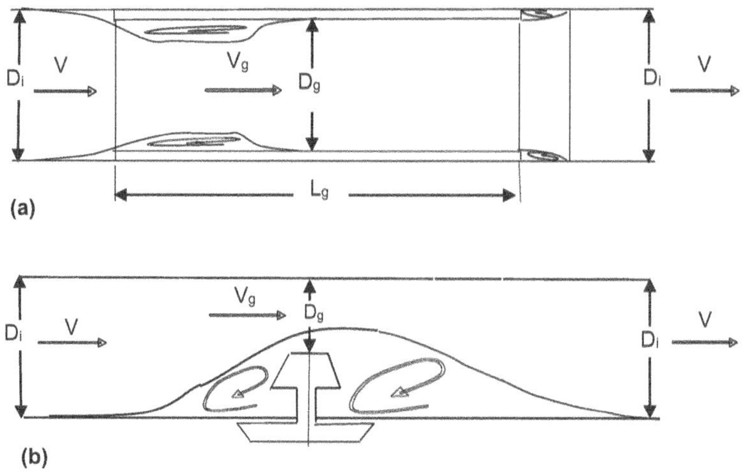

Fig. 5.1 Flow variation (sudden contraction and subsequent enlargement) and hydraulic variables along lateral with **a**: Integrated in-line emitter, **b**: on-line emitter

Recently, the author [16, 23, 25] presented simple analytical procedures for hydraulic design of trickle laterals which takes into consideration the effect of minor head losses, expressing the amount of minor head losses as a fraction of the kinetic head, as well as the effect of the emitter outflow non-uniformity, the kinetic head change, the number of emitters, and different uniform line slopes, on the lateral hydraulic computations. For any desired uniformity level, the analytical procedure gives one an opportunity to evaluate the influence of local energy loss on the pipe geometric characteristics (pipe size and length) and on the corresponding hydraulic variables (operating inlet and downstream end pressure heads and total energy losses) [17, 21, 22]. This study showed that in some design cases, neglecting minor losses may lead to erroneous designs of the lateral diameter and length [42–49].

This paper offers a simple mathematical model based on the stepwise procedure to accurately determine the pressure head profile along the lateral line. Essentially, the present work extends the earlier discussion [18] on the experimental analysis of local pressure losses along micro-irrigation laterals [9] to a systematic comparison of different kinds of in-line/on-line emitters and various design configurations. In this chapter, relative contributions of each of the energy loss components are calculated with the appropriate mathematical equation. This procedure is used to evaluate two examples with different types of in-line and on-line emitters; thus the results are compared with those of obtained from the recent analytical and experimental procedures, for all performed simulations.

Theoretical Development for Total Head (Energy) Loss Components

Minor Friction Losses Through Integrated In-Line Emitters

Integrated in-line emitters cause contraction of the flow path at the upstream connection between the emitter and the lateral pipe, and the expansion of the flow path immediately downstream from the emitter; thus, an additional minor friction loss must be considered [9]. Integrated in-line emitters have an inner diameter, D_g (m), smaller than the pipe inner diameter, D_i (m), and therefore the emitters determine higher frictional head losses due to the lower cross-section area. Emitter friction loss per unit emitter length, J_e (m/m), can also be evaluated by the Darcy-Weisbach friction formula:

$$J_e = \frac{f_e}{D_g}\frac{V_g^2}{2g} \tag{5.1}$$

where f_e = friction coefficient for the emitter flow; D_g = internal diameter of an integrated in-line emitter (m); V_g = flow velocity inside the emitter (ms^{-1}); and g = acceleration due to gravity (ms^{-2}).

The emitter friction loss for an individual in-line emitter, $h_{f(e)}$, can be evaluated from:

$$h_{f(e)} = J_e L_g \tag{5.2}$$

where L_g = longitudinal length of integrated in-line emitter (m).

Using the continuity equation, the following relationships between the pipe and the emitter flow velocity, V and V_g, can then be written:

$$V_g = V\left(\frac{D_i}{D_g}\right)^2 = \frac{4}{\pi}\frac{Q}{D_g^2} \tag{5.3}$$

where V = flow velocity in the pipe section (ms^{-1}); D_i = lateral pipe inner diameter (m); and Q = total lateral inflow rate (m^3s^{-1}).

For the interval of the Reynolds number, R, 2,000 < R < 36,000 for the friction coefficient, f, the Blasius equation is practically used to determine friction losses. For a large range of the Reynolds number, for the flow into the integrated in-line emitter the following expression can be used:

$$f_e = cR_g^{-0.25} \tag{5.4}$$

where c = coefficient which can be assumed to equal 0.316 and 0.302 [3].

Where R_g = Reynolds number for the flow occurred in the integrated in-line emitter which can be expressed as:

$$R_g = \frac{V_g D_g}{\nu} = \frac{4}{\pi \nu} \frac{Q}{D_g} \tag{5.5}$$

where ν = kinematic viscosity of water at standard temperature, $\nu = 1.01 \times 10^{-6}\ m^2s^{-1}$.

Substituting Eq. (5.5) into Eq. (5.4), then yields

$$f_e = c\left(\frac{\pi \nu}{4}\right)^{0.25}\left(\frac{D_g}{Q}\right)^{0.25} \tag{5.6}$$

Substituting Eqs. (5.3) and (5.6) into Eq. (5.1) then, into Eq. (5.2), the friction loss in the integrated in-line emitter along its length, L_g, can then be written:

$$h_{f(e)} = K\frac{Q^{1.75}}{D_g^{4.75}}L_g \tag{5.7}$$

where

$$K = c\frac{\upsilon^{0.25}}{2g}\left(\frac{4}{\pi}\right)^{1.75} \tag{5.8}$$

The value of the constant K is equal to 7.792 10^{-4} if the friction coefficient is determined by using the classical Blasius formula (c = 0.316) and 7.447 10^{-4} if the value of c is taken into account 0.302 [3].

Assuming N emitters are located from the lateral inlet (1st emitter) toward to the downstream closed end (Nth emitter), using Eq. (5.7), then the sum of friction losses along the integrated in-line emitters, $\Delta H_{f(e)}$, can be practically evaluated from the following mathematical expression

$$\Delta H_{f(e)} = \sum_{i=1}^{i=N} \Delta H_{f(e)i} = KL_g\frac{Q_n^{1.75}}{D_g^{4.75}}\sum_{i=1}^{i=N} i^{1.75} \tag{5.9}$$

where Q_n = nominal flow rate which assumed equal to the average value of the emitter outflow (Q_{av}) (Lh^{-1}); and i = an integer represents consecutive order of emitters. It can be assumed, the nominal flow rate is approximately equal to the average emitter outflow when the maximum allowable difference in the extreme pressure heads along the lateral is approximated by 10% of the average pressure head [11].

Local Energy Losses Due to Presence of Integrated In-Line/On-Line Emitters

The introduction of an integrated in-line or on-line emitter in a lateral causes a local pressure loss, λ, due to the obstruction of emitter connection into the pipe flow, that can be expressed as an "α" fraction of the kinetic head [14]:

$$\lambda = \alpha \frac{V^2}{2g} = \alpha \frac{8}{\pi^2 g} \frac{Q^2}{D_i^4} \tag{5.10}$$

where α = local loss coefficient due to both the contraction and the subsequent enlargement, can then be expressed as a function of the diameter ratio D_i/D_g, which is experimentally verified by the following expression [9, 10]:

$$\alpha = 0.056\left[\left(\frac{D_i}{D_g}\right)^{17.83} - 1\right] \tag{5.11}$$

The amount of local pressure losses due to the presence of integrated in-line and/or on-line emitters, ΔH_l, can be easily determined by the following mathematical expression (Provenzano et al. 2005):

$$\Delta H_l = \sum_{i=1}^{i=N-1} \Delta H_{li} = \alpha\left(\frac{8}{\pi^2 g}\right)\frac{Q_n^2}{D_i^4}\sum_{i=1}^{i=N-1} i^2 \tag{5.12}$$

Major Friction Losses Along the Lateral with Integrated In-Line/On-Line Emitters

Assuming integrated in-line emitters are located at an equal spacing, S then, sum of the friction losses along the length of the straight pipe sections between the consecutive in-line emitters, $(S–L_g)$, can be computed from the following expression:

$$\Delta H_{f(p)} = \sum_{i=1}^{i=N-1} \Delta H_{f(p)i} = K\frac{Q_n^{1.75}}{D_i^{4.75}}(S - L_g)\sum_{i=1}^{i=N-1} i^{1.75} \tag{5.13}$$

For the lateral with on-line emitters, Eq. (5.13) simply transforms to the following:

$$\Delta H_{f(p)} = \sum_{i=1}^{i=N-1} \Delta H_{f(p)i} = KS\frac{Q_n^{1.75}}{D_i^{4.75}}\sum_{i=1}^{i=N-1} i^{1.75} \tag{5.14}$$

Total Energy Losses Along the Lateral with Integrated In-Line and/or On-Line Emitters

Finally, the amount of total energy losses, ΔH_T, is a function of three energy loss components given by Eqs. (5.9), (5.12) and (5.13) or (5.14), finally, it can be approximated by the following expression:

$$\Delta H_T = \sum_{i=1}^{i=N} \Delta H_{Ti} = \Delta H_{fi} + \Delta H_{li} = \Delta H_{f(p)i} + \Delta H_{f(e)i} + \Delta H_{li} \tag{5.15}$$

Computational Applications

For Integrated In-Line Emitter Model

In order to check the practicability of the present procedure, we will try to compute total energy loss components using the simple formulations deduced in here. An *Excel* spreadsheet was set up to carry out the hydraulic calculations.

The analysis was realized by considering two types of integrated in-line emitters B and C which are characterized by the following experimental relationships given by [9] (please, see Table 5.1):

$$\textbf{B}\text{ model } (Siplast\ Tandom): \quad Q_{(Lh^{-1})} = 0.612 H_{(m)}^{0.537} \tag{5.16}$$

$$\textbf{C}\text{ model } (Rainbird\ Goccialin): \quad Q_{(Lh^{-1})} = 0.667 H_{(m)}^{0.514} \tag{5.17}$$

The analysis was conducted with the zero slope case and with standard commercial emitter spacing of 0.2, 0.4, 0.6, 0.8, 1.0, and 1.5 m. The specifications of both the emitter models are as follows:

For the **B** model:

Nominal flow rate of the emitter outflow (as indicated by the manufacturer), $Q_n = 2.1\ Lh^{-1} = 5.833\ 10^{-7}\ m^3s^{-1}$; nominal diameter and inner diameter of the lateral pipe, ND = 16 mm and $D_i = 13.29$ mm, respectively; inner diameter and longitudinal length of the integrated in-line emitter, $D_g = 11.51$ mm and $L_g = 67.89$ mm, respectively. The ratio of the pipe diameter to the emitter diameter, $D_i/D_g = 1.155$ and therefore the local loss coefficient from Eq. (5.11):

$$\alpha = 0.056\left[\left(\frac{13.29}{11.51}\right)^{17.83} - 1\right] = 0.671$$

For the **C** model:

$Q_n = 2.2\ Lh^{-1} = 6.111\ 10^{-7}\ m^3s^{-1}$; ND = 16 mm, $D_i = 13.46$ mm; $D_g = 12.14$ mm; and $L_g = 68.22$ mm. $D_i/D_g = 1.108$ and finally the local loss coefficient from Eq. (5.11), $\alpha = 0.297$, is computed. Further discussion on results for this application will be presented in the following section "*Results and Discussion*".

Table 5.1 Values of hydraulic characteristics for two kinds of the integrated in-line emitter models (B: *Siplast Tandem* and C: *Rainbird Goccialin*) for different emitter spacing (S) ranging from 0.2 to 1.5 m, and various number of emitters (N) for the desired (fixed) level of the Christiansen's uniformity coefficient, UC = 97%, for the present mathematical model

	B model (*Siplast Tandem*) ($\alpha = 0.671$)						C model (*Rainbird Goccialin*) ($\alpha = 0.297$)					
S (m)	N (–)	L_e (m)	$Q_{in} = \sum_{i=1}^{i=N} q_i$ (Lh^{-1})	$H_{in} = H_{(1)}$ (m)	$H_{(N)}$ (m)	ΔH_T (m)	N (–)	L_e (m)	$Q_{in} = \sum_{i=1}^{i=N} q_i$ (Lh^{-1})	$H_{in} = H_{(1)}$ (m)	$H_{(N)}$ (m)	ΔH_T (m)
0.2	172	34.2	356.2	10.231	8.165	2.066	188	37.4	402.3	10.154	8.305	1.849
0.4	153	60.8	316.4	10.203	8.165	2.038	159	63.2	339.3	10.094	8.305	1.789
0.6	140	83.4	289.3	10.190	8.165	2.025	143	85.2	305.4	10.111	8.305	1.806
0.8	130	103.2	268.4	10.169	8.165	2.004	132	104.8	282.1	10.134	8.305	1.829
1.0	123	122.0	254.1	10.191	8.165	2.026	123	122.0	262.7	10.125	8.305	1.820
1.5	109	162.0	224.9	10.166	8.165	2.001	109	162.0	233.3	10.174	8.305	1.869

N = total number of emitters along the lateral line; $L_e = (N-1)\,S$ = length of lateral between the first and last emitters; Q_{in} = total inflow rate which is accumulated by all emitter outflows; $H_{in} = H_{(1)}$ = pressure head at the first emitter in lateral inlet; $H_{(N)}$ = downstream closed end pressure headat the last emitter and; ΔH_T = the amount of total energy (head) losses which consist of total friction and local losses

For On-Line Emitter Model

Determine the percentages of total energy losses for two on-line emitter models (labyrinth and orifice-vortex) by considering the horizontal trickle lateral, for the following input parameters. Inner diameter of the lateral pipe is 13.4 mm (D_i = 13.4 mm), the emitter spacing, S, values are 0.3, 0.6, 0.9 and 1.2 m, respectively, and the overall Christiansen uniformity coefficient of the system is U_C = 98.8%. The following relationships with local loss coefficients, α, for both emitter models are experimentally verified as follows [8]:

Labyrinth on-line emitter model [$Q_{av} = 4.1\ Lh^{-1}$; $H_{av} = 9.65$ m; $\alpha = 0.34$]:

$$Q^{-1}_{(Lh\ )} = 1.32\,H^{0.5}_{(m)} \tag{5.18}$$

Orifice-vortex on-line emitter model [$Q_{av} = 4.2\ Lh^{-1}$; $H_{av} = 10.34$ m; $\alpha = 0.67$]:

$$Q^{-1}_{(Lh)} = 1.65\,H^{0.4}_{(m)} \tag{5.19}$$

Detailed analysis of results and discussion on this application will be presented in the following section "*Results and Discussion*".

Implementation of Backward Stepwise Procedure

The numerical stepwise procedure in the backward form based on the present mathematical model can be implemented for the data given in the present applications, starting from the downstream closed end (Nth emitter), toward to the lateral inlet (1st emitter).

First, in accordance with the first application (for integrated in-line emitter model), the following calculation steps are applied, respectively:

1. The outflow of the Nth emitter as a function of the corresponding pressure head, was assumed to equal 90% of that corresponding to the nominal discharge.

   The outflows at the downstream end for the B model: $q_N = 0.90 \times 2.1 = 1.89\ Lh^{-1}$ and, for the C model: $q_N = 0.90 \times 2.2 = 1.98\ Lh^{-1}$ and the corresponding pressure heads were computed from their outflow-pressure head relationships;
2. The friction loss along both the Nth emitter, $\Delta H_{f(e)\,(N)}$, by Eq. (5.9) and, the friction loss along the pipe segment between the consecutive Nth and (N–1)th emitters, $\Delta H_{f(p)\,(N-1)}$, by Eq. (5.13) then, the local loss due to insertion of the (N–1)th emitter, $\Delta H_{l\,(N-1)}$, by Eq. (5.12), were individually evaluated;
3. The pressure head for the (N–1)th emitter, as the pressure head for the Nth emitter plus the sum of total head losses between the two consecutive emitters then, the corresponding outflow for the (N–1)th emitter, $q_{\,(N-1)}$ were determined, respectively.
4. Using the continuity equation, the lateral discharge along the pipe segment between the consecutive Nth and (N–1)th emitters was computed by assuming the residual lateral flow rate at the downstream closed end from the last emitter is equal to zero, $Q_{(N)} = q_{(N)}$ and; $Q_{(N-1)} = Q_{(N)} + q_{(N-1)}$. Finally, total lateral inflow rate, Q_{in}, must be equal to the sum of whole emitter outflows, $Q_{in} = \sum_{i=1}^{i=N} q_i$ [46–54].
5. The uniformity level of the system was evaluated with the well-known Christiansen's uniformity coefficient, $UC = 1 - \left(1/NQ_n\right) \sum_{i=1}^{i=N} |q_i - Q_n|$. Noting that in the UC formula, the nominal discharge is taken into consideration as the average emitter outflow [24, 35, 40, 42–45]. Finally, the computation steps were improved to keep to the right the fixed value of UC = 97%. For each computation step, the uniformity coefficient UC was evaluated and then, the steps 1, 2, 3 and 4, were repeated until the desired value of UC = 97% was finally reached.

Results and Discussion

On Application-I (for Integrated In-Line Emitter Model)

In accordance with the first application (Application-I), Tables 5.1 and 5.2 synthesize the complete results for the main hydraulic variables regarding two kinds of integrated in-line emitter models **B** ($\alpha = 0.671$) and C ($\alpha = 0.297$) for the desired level of uniformity, UC = 97%.

From Table 5.1 for both the emitter models, the values of total inflow rate, Q_{in} (4th and 10th columns), the inlet pressure head at the upstream end, $H_{in} = H_{(1)}$ (5th and 11th columns) and the pressure head at the downstream closed end, $H_{(N)}$ (6th and 12th columns), the amount of the total energy losses, ΔH_T (7th and 13th columns) are calculated, respectively, regarding with different emitter spacing, S ranging from 0.2 to 1.5 m(1st column), various total number of emitters, N (2nd and 8th columns) and total length of the lateral line, L_e (3rd and 9th columns).

As indicated in Table 5.1, as the emitter spacing (S) increases and the total number of emitters (N) decreases, the total length of the lateral line (L_e) increases, in relation to the expression $L_e = (N-1)$ S, for both the emitter models. The total inflow rate (Q_{in}) decreases with decreasing in the total amount of emitter outflows ($Q_{in} = q_1 + q_2 + \ldots + q_N$)]. For both the emitter models, the inlet pressure head ($H_{in} = H_1$) yields similar values around 10 m, except for a little deviation.

It is clearly observed from this table [regarding the data sets for S = 1.0 m, N = 123 L_e = 122.0 m and for S = 1.5 m, N = 109, L_e = 162.0 m], the values of ΔH_T for the **B** model ($\alpha = 0.671$) with respect to the **C** model ($\alpha = 0.297$) increases with increasing the amount of local losses caused by the higher local loss coefficient, as also evaluated in Eq. (5.10). Regarding with different emitter spacing the ΔH_T yields fixed values; for instance, for the **B** model the ΔH_T yields values varying from 2.0 to 2.07 m, and for the **C** model varying from 1.79 to 1.87 m.

Regarding both the emitter models **B** and **C**, and different emitter spacing, S, various total number of emitters and total length of the lateral line, L_e, three components of total energy losses (ΔH_l, $\Delta H_{f(p)}$ and $\Delta H_{f(e)}$) are computed then the results synthesized, in Table 5.2.

In Table 5.2 first, the total energy losses (ΔH_T) are divided by the total friction losses (ΔH_f) and total emitter local losses (ΔH_l), and then, the total friction losses (ΔH_f) are subdivided by the major pipe friction losses ($\Delta H_{f(p)}$) and minor emitter friction losses ($\Delta H_{f(e)}$). Accordingly, the ratio of emitter local losses to the emitter friction losses [$\Phi_{e(l,f)} = \Delta H_l / \Delta H_{f(e)}$], and the emitter local losses as percentage of total energy losses [$\Phi_l(\%) = \Delta H_l / \Delta H_T$] are computed from the present mathematical model, and the results are compared with the recent experimental analysis [9].

Table 5.2 Relative contribution of three energy loss components [ΔH_l, $\Delta H_{f(p)}$, $\Delta H_{f(e)}$] to the amount of total energy losses (ΔH_T) with the amount of local losses as percentage of total energy losses [Φ_l] and the range of emitter local losses to the emitter friction losses [$\Phi_{e(l,f)}$], regarding two kinds of the integrated in-line emitter models (B: *Siplast Tandem* and C: *Rainbird Goccialin*), for different emitter spacing (S) ranging from 0.2–1.5 m and various number of emitters (N) from to 109 to 188, for the desired level of the Christiansen's uniformity coefficient, UC = 97%, according to the present mathematical model and the recent experimental procedure

	Present mathematical model								Recent experimental analysis						
S (m)	N (–)	L_e (m)	ΔH_T (m)		ΔH_f (m)		$\Phi_{e(l,f)} = \Delta H_l/\Delta H_{f(e)}$ (–)	$\Phi_l = \Delta H_l/\Delta H_T$ (%)	ΔH_T	ΔH_f (m)				$\Phi_l = \Delta H_l/\Delta H_T$ (%)
			ΔH_f (m)	ΔH_l (m)	$\Delta H_{f(p)}$ (m)	$\Delta H_{f(e)}$ (m)			ΔH_f (m)	ΔH_l (m)	$\Delta H_{f(p)}$ (m)	$\Delta H_{f(e)}$ (m)	$\Phi_{e(l,f)} = \Delta H_l/\Delta H_{f(e)}$ (–)	
B model (*siplast tandem*) ($\alpha = 0.671$)														
0.2	172	34.2	1.031	1.035	0.511	0.520	1.99	50.1	0.969	0.95	0.482	0.487	1.95	49.5
0.4	153	60.8	1.309	0.729	0.932	0.377	1.93	35.8	1.231	0.67	0.878	0.353	1.90	35.2
0.6	140	83.4	1.467	0.559	1.171	0.296	1.89	27.6	1.38	0.514	1.103	0.277	1.85	27.1
0.8	130	103	1.556	0.448	1.315	0.241	1.86	22.4	1.465	0.411	1.239	0.226	1.82	21.9
1.0	123	122	1.645	0.379	1.438	0.207	1.83	18.7	1.550	0.349	1.356	0.194	1.80	18.4
1.5	109	162	1.737	0.265	1.588	0.149	1.78	13.2	1.636	0.243	1.496	0.140	1.74	12.9
C model (*rainbird goccialin*) ($\alpha = 0.297$)														
0.2	188	37.4	1.226	0.623	0.664	0.562	1.11	33.7	1.156	0.573	0.626	0.530	1.08	33.1
0.4	159	63.2	1.412	0.377	1.057	0.355	1.06	21.1	1.33	0.347	0.996	0.334	1.04	20.7
0.6	143	85.2	1.531	0.275	1.266	0.265	1.04	15.2	1.444	0.253	1.194	0.250	1.01	14.9
0.8	132	105	1.612	0.216	1.399	0.213	1.01	11.8	1.521	0.199	1.320	0.201	0.99	11.6
1.0	123	122	1.644	0.175	1.469	0.175	1.0	9.6	1.551	0.161	1.385	0.166	0.97	9.4

(continued)

Table 5.2 (continued)

	Present mathematical model							Recent experimental analysis						
S (m)	N (−)	L_e (m)	ΔH_T (m)		ΔH_f (m)		$\Phi_{e(l,f)} = \Delta H_l/\Delta H_{f(e)}$ (−)	$\Phi_l = \Delta H_l/\Delta H_T$ (%)	ΔH_T	ΔH_f (m)				$\Phi_l = \Delta H_l/\Delta H_T$ (%)
			ΔH_f (m)	ΔH_l (m)	$\Delta H_{f(p)}$ (m)	$\Delta H_{f(e)}$ (m)			ΔH_f (m)	ΔH_l (m)	$\Delta H_{f(p)}$ (m)	$\Delta H_{f(e)}$ (m)	$\Phi_{e(l,f)} = \Delta H_l/\Delta H_{f(e)}$ (−)	
1.5	109	162	1.747	0.122	1.621	0.126	0.97	6.5	1.649	0.112	1.530	0.119	0.94	6.4

From this table, the analysis for three energy loss components reveals that, the emitter local losses (ΔH_l) and the emitter minor friction losses ($\Delta H_{f(e)}$) decrease due to decreasing in the total number of emitters (N) [i.e, with increasing in the emitter spacing (S)], whereas the pipe friction losses ($\Delta H_{f(p)}$) [i.e, the total friction losses (ΔH_f)] increase due to increasing in the total length of the lateral line (L_e). Thus, the amount of total energy losses (ΔH_T) does not change more (please see, 7th and 13th columns, in Table 5.1), since the total increasing in the amount of ΔH_f is just balanced with the total decreasing in the amount of ΔH_l.

In Table 5.2, from the comparison of the energy loss parameters [4th, 5th, 6th and 7th columns], the following finding can be observed. For both the emitter models **B** and **C**, as the total number of emitters (N) decreases and the total length of the lateral line (L_e) increases, the discrepancy between the amount of the ΔH_f and the ΔH_l, and between the amount of the $\Delta H_{f(p)}$ and the $\Delta H_{f(e)}$ more increases. However, as an exceptional case for the B emitter model, if the smallest value of the emitter spacing (S = 0.2 m) is regarded (N = 172 and L_e = 34.2 m), the amount of emitter local losses (ΔH_l = 1.035 m) is nearly identical to the amount of total friction losses (ΔH_f = 1.031 m); similarly, the amount of emitter friction losses ($\Delta H_{f(e)}$ = 0.52 m) closely follows the amount of pipe friction losses ($\Delta H_{f(p)}$ = 0.511 m). For S = 0.2 m, these findings are also confirmed by the recent experimental work as follows: ΔH_f = 0.969 m with ΔH_l = 0.95 m, and $\Delta H_{f(p)}$ = 0.482 m with $\Delta H_{f(e)}$ = 0.487 m, respectively.

For the **B** emitter model, the ratio of emitter local losses to the emitter friction losses [8th column] generally yields around 2.0 (for S = 0.2 m, $\Phi_{e(l,f)}$ = 1.99], and little decreases to 1.78 for S = 1.5 m. The emitter local losses as the percentage of total energy losses [9th column] reach to the highest range of 50% [for S = 0.2 m, Φ_l = 50.1%], and rapidly decreases to 13.2% (for S = 1.5 m). The recent experimental work gives similar results as follows: $\Phi_{e(l,f)}$ values are 1.95 for S = 0.2 m and 1.74 for S = 1.5 m; and Φ_l values are 49.5(%) for S = 0.2 m and 12.9% for S = 1.5 m.

For the **C** emitter model with respect to the **B** model (regarding with 4–7th columns) for the smallest value of emitter spacing, S = 0.2 m, the amount of total friction losses [ΔH_f = 1.226 m] is approximately equal to two times of the amount of emitter local losses [ΔH_l = 0.623 m], since the amount of emitter local losses decrease in proportion to the value of the local loss coefficient for the **C** model (α = 0.297) which is approximately half of the value for the **B** model (α = 0.671). As also previously pointed out for the **B** emitter model (for S = 0.2 m), the amount of emitter friction losses [$\Delta H_{f(e)}$ = 0.562 m] nearly approaches to the amount of pipe friction losses [$\Delta H_{f(p)}$ = 0.664 m], for the **C** emitter model, as well. For S = 0.2 m, these findings are also justified with those of obtained in the recent experimental work as follows: ΔH_f = 1.156 m with ΔH_l = 0.573 m and $\Delta H_{f(p)}$ = 0.626 m with $\Delta H_{f(e)}$ = 0.53 m, respectively.

For the **C** emitter model, the ratio of emitter local losses to the emitter friction losses [8th column] generally yields around 1.0 [for S = 0.2 m, $\Phi_{e(l,f)}$ = 1.11] and little decreases to 0.97 for S = 1.5 m. The emitter local losses as the percentage of total energy losses [9th column] approach to the range of 34% [for S = 0.2 m, Φ_l = 33.7%], and rapidly decreases to 6.5% for S = 1.5 m. Hence, there is a good agreement with those of the experimental analysis as: $\Phi_{e(l,f)}$ values are 1.08 for S = 0.2 ma nd 0.94 for S = 1.5 m; and Φ_l values are 33.1(%) for S = 0.2 m and 6.4% for S = 1.5 m.

As a remarkable result from this table [regarding the data set given for S = 1.0 m (N = 123, L_e = 122.0 m) and for S = 1.5 m (N = 109, L_e = 162.0 m)] for both the emitter models (**B** and **C**), the amount of pipe friction losses for the **C** model [$\Delta H_{f(p)}$ = 1.47 m and 1.62 m] is higher than those of the **B** model [$\Delta H_{f(p)}$ = 1.44 m and 1.59 m]. Because, the **C** model has a higher nominal flow rate (Q_n = 2.2 Lh^{-1}) with respect to the **B** model (Q_n = 2.1 Lh^{-1}) even if the **C** model has higher values of the emitter length (L_g) and inner lateral diameter (D_i) than those of the B model, as concluded in Eq. (5.13).

For both the emitter models (**B** and **C**) [regarding the data set for S = 1.0 m (N = 123, L_e = 122.0 m) and for S = 1.5 m (N = 109, L_e = 162.0 m)], the amount of emitter friction losses for the **B** model [$\Delta H_{f(e)}$ = 0.207 m and 0.149 m] is higher than those of the **C** model [$\Delta H_{f(e)}$ = 0.175 m and 0.126 m]. Because, since the **B** model has a smaller inner emitter diameter (D_g = 11.51 mm) with respect to the **C** model (D_g = 12.14 mm) even if the **B** model has smaller values of the emitter length (L_g) and nominal flow rate (Q_n) than those of the **B** model, as concluded in Eq. (5.9).

For both the emitter models (**B** and **C**), to illustrate the variation of three energy loss components to each other, two figures were also prepared and represented by Figs. 5.2 and 5.3, respectively. For the **B** model of integrated-in-line emitter, Fig. 5.2 illustrates variation of total friction (ΔH_f) [premier axis of y] and emitter local (ΔH_l) losses [secondary axis of y] regarding with different values of emitter spacing (S) [premier axis of x], and various number of emitters (N) [secondary axis of x], according to both the mathematical (straight bold line) and experimental (dotted line) procedures. For the **C** model of integrated- in-line emitter, Fig. 5.3 illustrates variation of total friction loss components due to pipe ($\Delta H_{f(p)}$) and due to emitter ($\Delta H_{f(e)}$) with regarding different values of emitter spacing (S) and various number of emitters (N), according to both the mathematical (straight line) and experimental (dotted line) procedures. Figures 5.2 and 5.3 also reveal that, a good justification between the results of both the mathematical and experimental procedures is observed, for all performed simulations.

On Application-II (for On-Line Emitter Model)

The same calculation steps for the backward stepwise procedure clarified-above are repeated (from 1st to 5th) using the related formulations given only for on-line emitters [Eqs. (5.12), (5.14), (5.15), (5.18) and (5.19)] for the overall desired uniformity level, UC = 98.8%.

Table 5.3 synthesizes the complete results for the main hydraulic variables for two kinds of on-line emitter models [labyrinth (α = 0.34) and orifice-vortex (α = 0.67)]

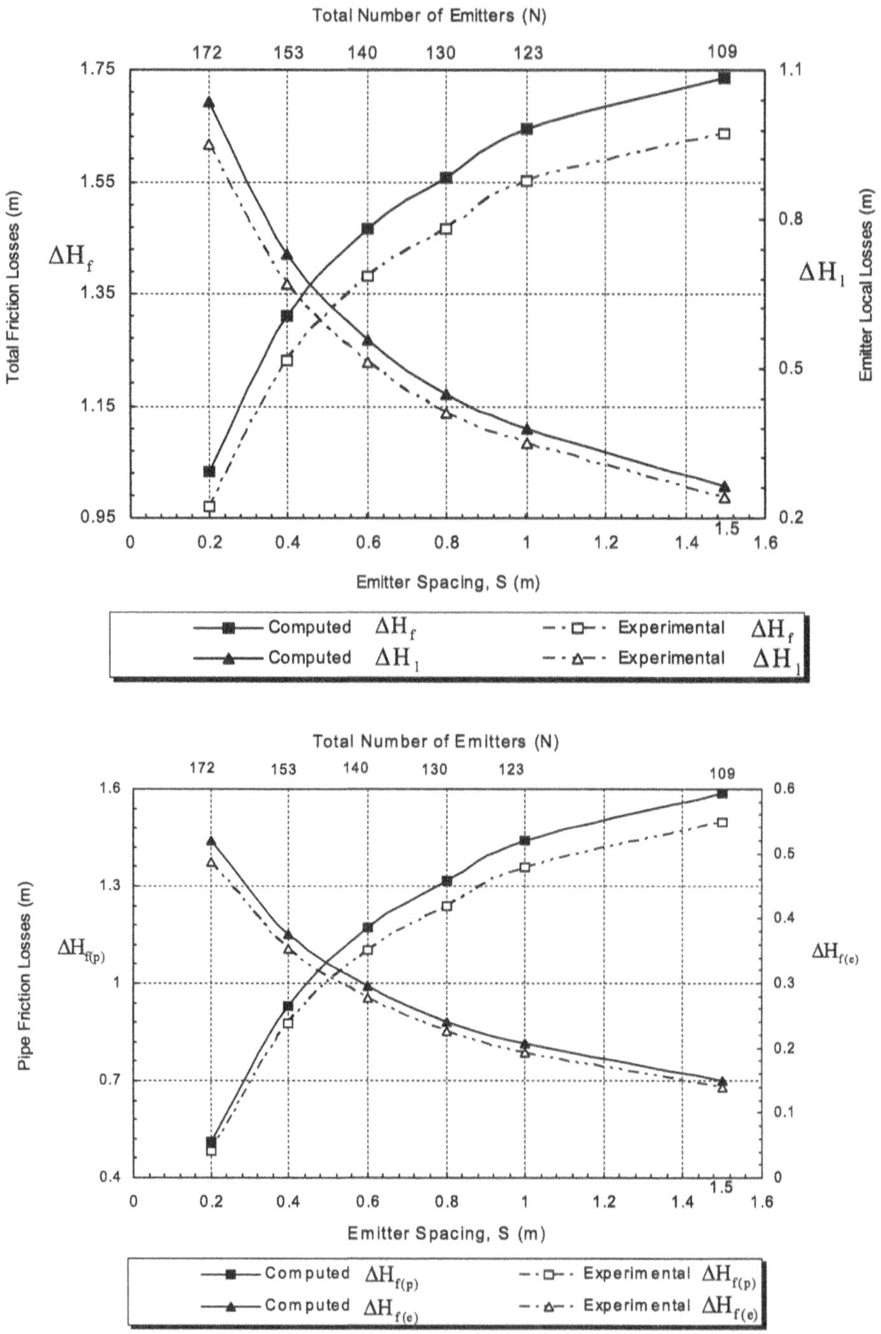

Fig. 5.2 For the B model (*Siplast Tandem*) of integrated-in-line emitter, variation of total friction [ΔH_f] and local [ΔH_l] losses versus different values of emitter spacing (S) varying from 0.2–1.5 m, and various number of emitters (N), according to both the mathematical and experimental procedures

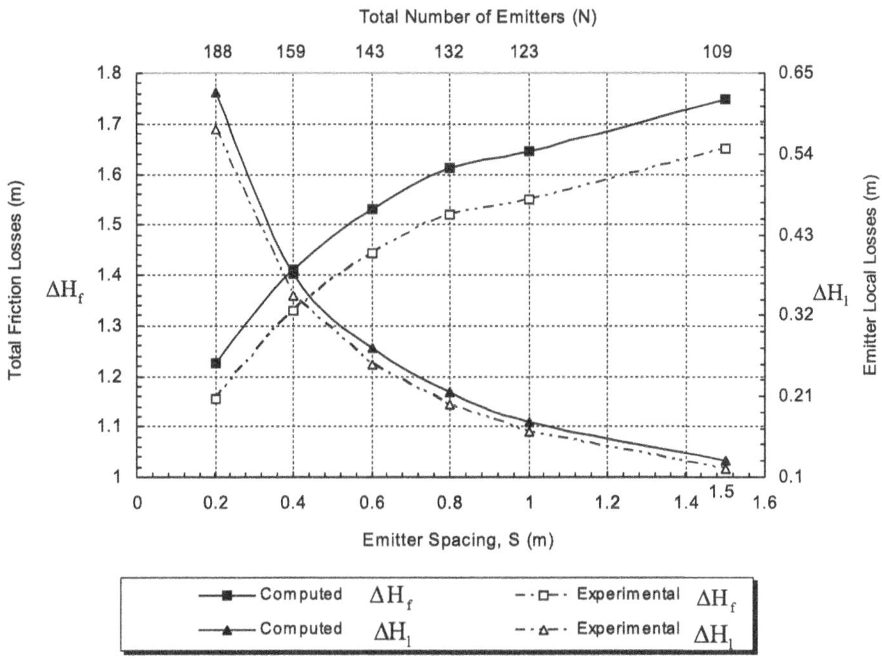

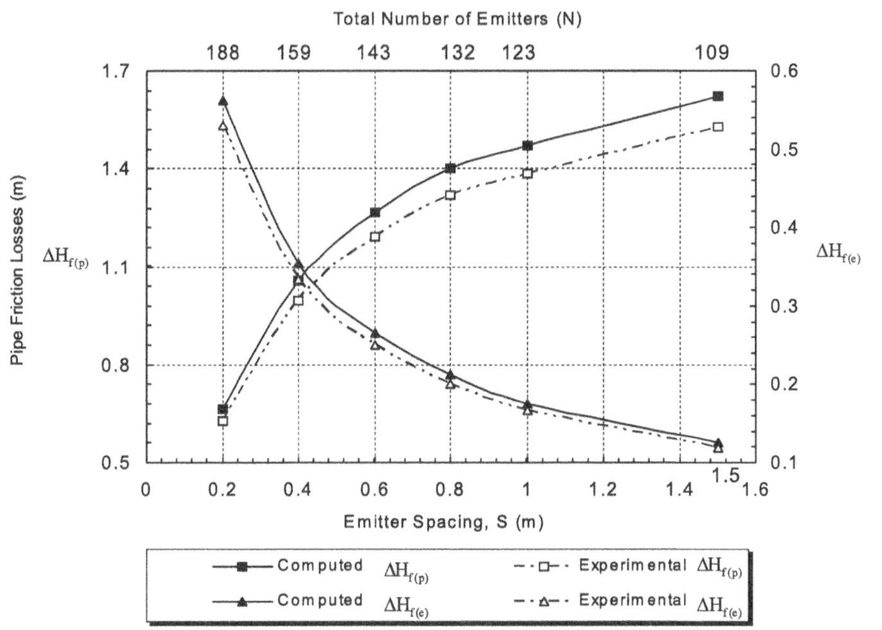

Fig. 5.3 For the C model (*Rainbird Goccialin*) of integrated-in-line emitter, variation of total friction loss components due to pipe [$\Delta H_{f(p)}$] and due to emitter [$\Delta H_{f(e)}$] versus different values of emitter spacing (S) varying from 0.2–1.5 m, and various number of emitters (N), according to both the mathematical and experimental procedures

in comparison with those of obtained from the previous analytical procedure [29]. In this table, for both the emitter models, two energy loss components (ΔH_f and ΔH_l) [4–8th and 5–9th columns], the ratio of total pipe friction losses to the emitter local losses, $\Phi_f = \Delta H_f/\Delta H_l$ [6 and 10th columns], and the emitter local losses as percentage of total energy losses, $\Phi_l = \Delta H_l/\Delta H_T$ [7 and 11th columns] regarding with different emitter spacing, S, total number of emitters, N, and total length of the lateral line, L_e, are presented, respectively.

Regarding the values presented in Table 5.3, the following remarks can be observed. First, the values of emitter local losses for the labyrinth on-line emitter model [$\Delta H_l =$ 0.598, 0.335, 0.241 and 0.192 m] are approximately half of the values for the orifice-vortex on-line emitter model [$\Delta H_l = 1.179$, 0.657, 0.459 and 0.374 m] since the emitter local losses proportionally increase with increasing the value of local loss coefficient, α (from 0.34–0.67).

Practically, as the length of the lateral increases and the number of emitters decreases with increasing in emitter spacing, the amount of emitter local losses decrease whereas the pipe friction losses increase; so the discrepancy between the pipe friction losses and the emitter local losses increases with increasing in the emitter spacing [1st, 2nd, 3rd, 4–8th and 5–9th columns].

For the labyrinth on-line emitter model ($\alpha = 0.34$), if the smallest value of the emitter spacing, S = 0.3 m, is selected, the amount of pipe friction losses is approximately equal to two times of the amount of emitter local losses whereas for the orifice-vortex on-line emitter model ($\alpha = 0.67$), the amount of pipe friction losses is approximately equal to the amount of emitter local losses.

Accordingly, for the labyrinth on-line emitter, the ratio of the friction losses to the emitter local losses (Φ_f) yields around 2, 4, 6 and 8 whereas it yields around 1, 2, 3 and 4, for the orifice-vortex on-line emitter, for the emitter spacing, S = 0.3, 0.6, 0.9 and 1.2 m, respectively. Moreover, the amount of emitter local losses, Φ_l (%) expressed as a percentage of the total energy losses, decreases with increasing emitter spacing, then its values can reach 32.9 and 50.1% (33.5 and 50% according to the previous analytical procedure) with respect to the labyrinth and orifice-vortex on-line emitters, respectively. It can be concluded from this comparison, a good justification is observed between the results obtained from the present mathematical and the previous analytical procedures.

Table 5.3 Relative contribution of two energy loss components [ΔH_f and ΔH_l] to the amount of total energy losses [ΔH_T] with the range of total friction losses to the total emitter local losses [Φ_f], and the amount of local losses as percentage of total energy losses [Φ_l], regarding two kinds of the on-line emitter models (*Labyrinth and orifice-vortex*), for different emitter spacing (S) ranging from 0.3–1.2 m and various number of emitters (N) from to 75–110, for the desired level of the Christiansen's uniformity coefficient, UC = 98.8%, according to the present mathematical model and the previous analytical procedure

			Present mathematical model				Previous analytical method [29]			
S (m)	N (–)	L_e (m)	ΔH_T (m)		$\Phi_f = \Delta H_f/\Delta H_l$ (–)	$\Phi_l = \Delta H_l/\Delta H_T$ (%)	ΔH_T (m)		$\Phi_f = \Delta H_f/\Delta H_l$ (–)	$\Phi_l = \Delta H_l/\Delta H_T$ (%)
			ΔH_f (m)	ΔH_l (m)			ΔH_f (m)	ΔH_l (m)		
Labyrinth on-line emitter ($\alpha = 0.34$)										
0.3	110	32.70	1.215	0.598	2.03 ≅ 2.0	32.9	1.108	0.559	1.98 ≅ 2.0	33.5
0.6	90	53.40	1.322	0.335	3.95 ≅ 4.0	20.2	1.277	0.307	4.16 ≅ 4.0	19.4
0.9	80	71.10	1.421	0.241	5.89 ≅ 6.0	14.5	1.386	0.216	6.42 ≅ 6.0	13.5
1.2	75	88.80	1.578	0.192	8.22 ≅ 8.0	10.9	1.548	0.178	8.70 ≅ 9.0	10.3
Orifice-vortex on-line emiter ($\alpha = 0.67$)										
0.3	110	32.70	1.172	1.179	0.99 ≅ 1.0	50.1	1.155	1.156	0.99 ≅ 1.0	50.0
0.6	90	53.40	1.343	0.657	2.04 ≅ 2.0	32.9	1.332	0.635	2.10 ≅ 2.0	32.3
0.9	80	71.10	1.462	0.459	3.18 ≅ 3.0	23.9	1.446	0.447	3.23 ≅ 3.0	23.6
1.2	75	88.80	1.625	0.374	4.34 ≅ 4.0	18.9	1.614	0.368	4.38 ≅ 4.0	18.6

$\Phi_f = \Delta H_f/\Delta H_l$: the ratio of total friction losses to the total emitter local losses; $\Phi_l = \Delta H_l/\Delta H_T$: the amount of total emitter local losses as percentage of total energy losses

Concluding Remarks

In this chapter a simplified mathematical model based on the backward stepwise for three energy loss components (local and friction losses due to emitters and pipe friction losses) to finally evaluate total energy losses along a trickle lateral line with integrated in-line and/or on-line emitters, is presented. The proposed Eqs. (5.9), (5.13) or (5.14) allow one to evaluate the minor friction losses through the path of an integrated in-line emitter, and the major friction losses along the pipe, and Eq. (5.12) allows one to evaluate the local losses attributable to the emitters' connection. Hence, an approximate combination formulation [Eq. (5.15)] which incorporates three energy loss components allows one to accurately evaluate total energy losses at the end of the lateral line.

The present technique is applied on two numerical applications covering different types of integrated in-line and on-line emitters to demonstrate its practicability and validity with respect to the recent literature. Examination of results for applications confirmed that, the proposed technique is efficient in all design cases examined, and justifies with the results reported in the recent analytical and experimental procedures. Based on the present assessment regarding different kinds of integrated in-line and on-line emitters, the following remarks can be underlined:

1. First, the amount of emitter local losses (ΔH_l) increases with increasing in the local loss coefficient, α.
2. Reducing the length of the lateral (L_e) and increasing the number of emitters (N) with decreasing the emitter spacing (S), the amount of local (ΔH_l) [and friction, $\Delta H_{f(e)}$] losses due to presence of emitters increase whereas the pipe friction losses ($\Delta H_{f(p)}$) (ie., total friction losses, ΔH_f) decrease. For both the integrated in-line emitter models **B** and **C** (Table 5.2), the discrepancy between the amount of the ΔH_f and the ΔH_l; and between the amount of the $\Delta H_{f(p)}$ and the $\Delta H_{f(e)}$ more increases as the total number of emitters (N) decreases and the total length of the lateral line (L_e) increases. Therefore, the amount of total energy losses (ΔH_T) (which is combination of total friction and emitter local losses) does not change more, except for a little deviation (Table 5.1), since the total increasing in the amount of total friction losses (ΔH_f) is just balanced with the total decreasing in the amount of emitter local losses (ΔH_l).
3. For both the **B** and **C** integrated in-line emitter models (Table 5.1), and regarding different values of the emitter spacing and total number of emitters the inlet pressure head ($H_{in} = H_1$) yields similar values around 10 m, since the amount of total energy losses (ΔH_T) yields fixed values, in all design cases examined.
4. The amount of emitter local losses, Φ_l, expressed as the percentage of total energy losses (ΔH_T) increases with decreasing in the emitter spacing. As an exceptional case for the smallest value, S = 0.2 m, the summation of emitter local losses nearly approach to half of the amount of the total energy losses [$\Phi_l = \Delta H_l / \Delta H_T \cong 50\%$].

5. For both the integrated in-line [B-siplast tandem: $\alpha = 0.671$] and on-line [orifice vortex: $\alpha = 0.67$] emitters (Tables 5.2 and 5.3), and regarding the data set for the smallest value of the emitter spacing (S = 0.2 or 0.3 m), the summation of emitter local losses nearly identical to the amount of total friction losses (due to pipe and emitters) [$\Delta H_f \cong \Delta H_l$].
6. For both the **B** and **C** integrated in-line emitter models (Table 5.2), and regarding the data set for the smallest value of the emitter spacing (S = 0.2 m), the summation of minor friction losses due to in-line emitters nearly identical to the amount of major friction losses along the pipe sections between successive in-line emitters [$\Delta H_{f(e)} \cong \Delta H_{f(p)}$].
7. Regarding the data set given for different emitter spacing (from 0.2 to 1.5 m) (Table 5.2), if the **B** model ($\alpha = 0.671$) is selected the summation of emitter local losses is approximately equal to two times of the amount of emitter minor friction losses [$\Delta H_l \cong 2 \times \Delta H_{f(e)}$], whereas for the **C** model ($\alpha = 0.297$) the summation of emitter local losses is nearly identical to the summation of minor friction losses due to in-line emitters [$\Delta H_l \cong \Delta H_{f(e)}$].

Acknowledgements The senior author wants to express his acknowledges to Prof. Vijay. P. Singh at Texas A&M University, Emeritus Prof. Necati Ağıralioğlu from International Antalya Science University, and Distinguished Emeritus Professor Zekai Şen from Istanbul Medipol University, for their highly encouragement on preparing this book, many scientific activities, projects and so on. The Scientific and Technological Research Council of Turkey (TUBITAK) is acknowledged for supporting by the fellowships and grant program during the studies in the Texas A&M University/BAEN. We also thank to publishers, potential readers and reviewers for providing constructive reviews and criticism on the content of the book, in advance.

Conflict of Interest The author/editor has no conflicts of interest, financial or otherwise, to declare.

References

1. A.I. Al-Amoud, Significance of energy losses due to emitter connections in trickle irrigation lines. J. Agric. Eng. Res. **60**(1), 1–5 (1995)
2. H. Anyoji, I.P. Wu, Statistical approach for drip lateral design. Trans. ASAE **30**(1), 187–192 (1987)
3. V. Bagarello, V. Ferro, G. Provenzano, D. Pumo, "Experimental study on flow resistance law for small diameter plastic pipes." J. Irrig. Drain. Eng., ASCE, 121(5), 313–316, 1995.
4. V. Bagarello, V. Ferro, G. Provenzano, D. Pumo, Evaluating pressure losses in drip irrigation lines. J. Irrig. Drain. Eng., ASCE. **123**(1), 1–7 (1997)
5. H.M. Hathoot, A.I. Al-Amoud, A.S. Al-Mesned, Design of trickle irrigation laterals considering emitter losses. Int. Comm. Irrig. Drain. (ICID) J. **49**(2), 1–14 (2000)

6. T.A. Howell, F.A. Barinas, Pressure losses across trickle irrigation fittings and emitters. Trans. ASAE **23**(4), 928–933 (1980)
7. L. Juana, L. Rodriguez-Sinobas, A. Losada, Determining minor head losses in drip irrigation laterals. I: methodology. J. Irrig. Drain. Eng. ASCE. **128**(6), 376–384 (2002a)
8. L. Juana, L. Rodriguez-Sinobas, A. Losada, Determining minor head losses in drip irrigation laterals. II: experimental study and validation. J. Irrig. Drain. Eng. ASCE. **128**(6), 385–396 (2002b)
9. G. Provenzano, D. Pumo, Experimental analysis of local pressure losses for microirrigation laterals. J. Irrig. Drain. Eng., ASCE. **130**(4), 318–324 (2004)
10. G. Provenzano, D. Pumo, P. Di Dio, Simplified procedure to evaluate head losses in drip irrigation laterals. J. Irrig. Drain. Eng., ASCE. **131**(6), 525–532 (2005)
11. G. Provenzano, D. Pumo, Closure to "Experimental analysis of local pressure losses for microirrigation laterals." By Giuseppe Provenzano and Domenico Pumo. J. Irrig. Drain. Engrg., ASCE **132**(2), 193–194 (2006)
12. E.J. Scaloppi, R.G. Allen, Hydraulics of irrigation laterals: comparative analysis. J. Irrig. Drain. Eng., ASCE. **119**(1), 91–115 (1993)
13. H.M. Hathoot, A.I. Al-Amoud, F.S. Mohammad, Analysis and design of trickle irrigation laterals. J. Irrig. Drain. Eng. ASCE. 119(5), 756−767 (1993)
14. R.W. Jeppson, *Analysis of flow in pipe networks*, 5th edn. (Ann Arbor Science, Ann Arbor, Mich, 1982)
15. G. Yıldırım, Analytical relationships for designing multiple outlets pipelines. J. Irrig. Drain. Eng. ASCE. **133**(2), 140−154 (2007a)
16. G. Yıldırım, An assessment of hydraulic design of trickle laterals considering effect of minor losses. Irrig. Drain. (ICID) J. **56**(4), 399–421 (2007b)
17. G. Yıldırım, Hydraulic analysis and direct design of multiple outlets pipelines laid on flat and sloping lands. J. Irrig. Drain. Eng. ASCE. **132**(6), 537–552 (2006a)
18. G. Yıldırım, Discussion of "Experimental analysis of local pressure losses for microirrigation laterals' by Provenzano and Pumo". J. Irrig. Drain. Engrg. ASCE. **132**(2), 189–192, (2006b)
19. G. Yıldırım, N. Ağıralioğlu, Discussion of "Determining minor losses in drip irrigation laterals. I: methodology." J. Irrig. Drain. Engrg. ASCE. **130**(3), 248–252 (2004c)
20. G. Yıldırım, N. Ağıralioğlu, Discussion of "Determining minor losses in drip irrigation laterals. II: experimental study and validation." J. Irrig. Drain. Engrg. ASCE. **130**(4), 344–346, (2004d)
21. G. Yıldırım, Comparative analysis of hydraulic calculation methods in design oh micro-irrigation laterals and linear solution of hydraulic analysis of tapered laterals PhD. Thesis. (Istanbul Technical University, Hydraulics Division, Istanbul, Turkey, 2007)
22. G. Yıldırım, *Computer-aided design of micro-irrigation laterals by using LatCAD* (Thesis, Istanbul Technical University, Hydraulics Division, Istanbul, Turkey, MSc, 2001)
23. G. Yıldırım, Total energy loss assessment for trickle lateral lines equipped with in-line and on-line emitters. Irrig. Sci. **28**(4), 341–352 (2010)
24. G. Yıldırım, Discussion of "Statistical analysis of non-pressure-compensating and pressure-compensating drip emitters" by H. Perea, J. Enciso-Medina, V. P. Singh, D. P. Dutta, and B. J. Lesikar. J. Irrig. Drain. Eng., ASCE (2016). https://doi.org/10.1061/(ASCE)IR.1943-4774.0000644
25. G. Yıldırım, Simplified procedure for hydraulic design of small-diameter plastic pipes. Irrig. Drain. (ICID J). **58**(2), 209–233 (2009a)
26. G. Yıldırım, An assessment for hydraulically efficient design of uniformly sloping submain, lines. Irrig. Drain. **64**(2), 157–168 (2015)
27. J.D. Valiantzas, Analytical approach for direct drip lateral hydraulic calculation. J. Irrig. Drain. Eng. ASCE **124**(6), 300–305 (1998)

28. J.D. Valiantzas, Continuous outflow variation along irrigation laterals: effect of the number of outlets. J. Irrig. Drain. Eng. ASCE. **128**(1), 34–42 (2002)
29. P. Vallesquino, P.L. Luque-Escamilla, New algorithm for hydraulic calculation in irrigation laterals. J. Irrig. Drain. Eng. ASCE. **127**(4): 254–260 (2001)
30. P. Vallesquino P.L. Luque-Escamilla, Equivalent friction factor method for hydraulic calculation in irrigation laterals. J. Irrig. Drain. Eng. ASCE. **128**(5), 278−286 (2002)
31. I.P. Wu, H.M. Gitlin, Energy gradient line for drip irrigation laterals. J. Irrig. Drain. Eng. ASCE **101**(4), 323–326 (1975)
32. I.P. Wu, Energy gradient line approach for direct hydraulic calculation in drip irrigation design. Irrig. Sci. **13**, 21–29 (1992)
33. I.P. Wu, An assessment of hydraulic design of microirrigation systems. Agric. Water Manag. **32**, 275–284 (1997)
34. I.P. Wu, R. Yue, Drip lateral design using energy gradient line approach. Trans. ASAE. **36**(2): 389–394, 10 (1993)
35. G. Yıldırım, O. Kose, New software (multi-flowCAD) development for sub-main line hydraulic design: computational algorithm, computer visualization and implementation. Arab. J. Sci. Eng. **41**, 1551–1560 (2016)
36. G. Yıldırım, Computer-based analysis of hydraulic design variables for uniformly sloping microirrigation system laterals. J. Irrig. Drain. Eng., ASCE. **141**(7), 04014079 (2015)
37. G. Yıldırım, V.P. Singh, A mathCAD procedure for commercial pipeline hydraulic design considering local energy losses. Adv. Eng. Softw. **41**(2), 489–496 (2010)
38. G. Yıldırım, Computer-based analysis of explicit approximations to the implicit colebrook-white equation in turbulent flow friction factor calculation. Adv. Eng. Softw. **40**(11), 1183–1190 (2009)
39. G. Yıldırım, M. Özger, Neuro-fuzzy approach in estimating Hazen-Williams Friction coefficient for small-diameter polyethylene pipes. Adv. Eng. Softw. **40**(8), 593–599 (2009)
40. M. Özger, G. Yıldırım, Determining turbulent flow friction coefficient using adaptive neuro-fuzzy computing technique. Adv. Eng. Softw. **40**(4), 281–287 (2009)
41. Y. Kang, S. Nishiyama, Analysis and design of microirrigation laterals. J. Irrig. Drain. Eng. ASCE **122**(2), 75–82 (1996)
42. G. Yıldırım, Discussion of "Field-scale assessment of uncertainties in drip irrigation lateral parameters" by Yeboah Gyasi-Agyei. J. Irrig. Drain. Eng. ASCE. **135**(2), 261–263 (2009b)
43. G. Yıldırım, Discussion of "Simplified method for sizing laterals with two or more diameters by Muluneh Yitayew." J. Irrig. Drain. Eng., ASCE (accepted for publication) (2009c)
44. G. Yıldırım, Determining operating "Inlet pressure head incorporating uniformity parameters for multioutlet plastic pipelines." J. Irrig. Drain. Eng., ASCE. **134**(3), 341–348 (2008a)
45. G. Yıldırım, Discussion of "Turbulent flow friction factor calculation using a mathematically exact alternative to the colebrook-white equation' by Sonnad and Goudar." J. Hydraulic Eng., ASCE. **134**(8), 1185–1186 (2008b)
46. G. Yıldırım, N. Ağıralioğlu, Closure to "Comparative analysis of hydraulic calculation methods in design of microirrigation laterals" by G. Yıldırım and N. Ağıralioğlu. J. Irrig. Drain. Eng., ASCE. **132**(1), 85–90 (2006c)
47. G. Yıldırım, N. Ağıralioğlu, Discussion of "Inlet pressure, energy cost, and economic design of tapered irrigation submains' by Valiantzas." J. Irrig. Drain. Engrg. ASCE. **131**(2), 220–224, (2005a)
48. G. Yıldırım, N. Ağıralioğlu, Discussion of "Explicit hydraulic design of microirrigation submain units with tapered manifold and laterals' by Valiantzas." J. Irrig. Drain. Engrg. ASCE. **131**(3), 299–300 (2005b)

49. G. Yıldırım, N. Ağıralioğlu Closure to "Linear solution for hydraulic analysis of tapered microirrigation laterals' by Yıldırım and Ağıralioğlu". J. Irrig. Drain. Eng., ASCE. **131**(5), 490–491, (2005c)
50. G. Yıldırım, N. Ağıralioğlu, Comparative analysis of hydraulic calculation methods in design of microirrigation laterals. J. Irrig. Drain. Eng. ASCE. **130**(3): 201–217 (2004a)
51. G. Yıldırım, N. Ağıralioğlu, Linear solution for hydraulic analysis of tapered microirrigation laterals. J. Irrig. Drain. Eng. ASCE, **130**(1), 78–87 (2004b)
52. G. Yıldırım, N. Ağıralioğlu, Discussion of "Equivalent friction factor method for hydraulic calculation in irrigation laterals." J. Irrig. Drain. Engrg. ASCE. **130**(4), 335–339 (2004e)
53. G. Yıldırım, N. Ağıralioğlu, Discussion of "New algorithm for hydraulic calculation in irrigation laterals". J. Irrig. Drain. Engrg. ASCE. **129**(2), 142–143 (2003a)
54. G. Yıldırım, N. Ağıralioğlu, Discussion of "Continuous outflow variation along irrigation laterals: effect of the number of outlets. J. Irrig. and Drain. Engrg., ASCE. **129**(5), 382–386 (2003b)

Index

G. Yıldırım, *Hydraulic Principles and Design Concepts for Submain Units with Multiple Outlet Pipelines*, Synthesis Lectures on Engineering, Science, and Technology, https://doi.org/10.1007/978-3-031-32495-6

SPRINGER NATURE

GPSR Compliance

The European Union's (EU) General Product Safety Regulation (GPSR) is a set of rules that requires consumer products to be safe and our obligations to ensure this.

If you have any concerns about our products, you can contact us on ProductSafety@springernature.com

In case Publisher is established outside the EU, the EU authorized representative is:

Springer Nature Customer Service Center GmbH
Europaplatz 3
69115 Heidelberg, Germany

The manufacturer's authorised representative in the EU is Springer Nature Customer Service Centre GmbH, Europaplatz 3, 69115 Heidelberg, Germany. If you have any concerns regarding our products, please contact ProductSafety@springernature.com

Printed and bound by CPI Group (UK) Ltd, Croydon, CR0 4YY

30/07/2026

02182037-0017